ENERGY AND CULTURE

Energy and Culture
Perspectives on the Power to Work

Edited by

BRENDAN DOOLEY
International University Bremen, Germany

ASHGATE

Published by
Ashgate Publishing Limited
Gower House
Croft Road
Aldershot
Hampshire GU11 3HR
England

Ashgate Publishing Company
Suite 420
101 Cherry Street
Burlington, VT 05401-4405
USA

Ashgate website: http://www.ashgate.com

British Library Cataloguing in Publication Data
Energy and culture : perspectives on the power to work. -
    (Ashgate studies in environmental policy and practice)
    1.Power resources - Social aspects 2.Energy development -
    Social aspects
    I. Dooley, Brendan Maurice, 1953-
    333.7'9

Library of Congress Cataloging-in-Publication Data
Energy and culture : perspectives on the power to work / edited by Brendan Dooley.
    p. cm. -- (Ashgate studies in environmental policy and practice)
    Includes index.
    ISBN 0-7546-4514-2
    1. Energy development--History. 2. Power resources--History. 3. Energy
consumption--History. 4. Energy policy--History. I. Dooley, Brendan Maurice, 1953-
II. Series.

    HD9502.A2E5386 2005
    333.79--dc22

                                                                                        2005022765

ISBN 0 7546 4514 2

Printed and bound in Great Britain by MPG Books Ltd. Bodmin, Cornwall.

Contents

List of Figures

List of Tables

List of Contributors

Bernard Beaudreau is Professor of Economics at the Université Laval, Quebec. His publications include *Energy and Organization: Growth and Distribution Reexamined*, Westview CT: 1998, and *Energy and the Rise and Fall of Political Economy*, Westview CT: 1999.

Giuliana Biagioli is Professor of History at the University of Pisa and director of the Centro di ricerca sulla storia dell'agricoltura e della società rurale. Her books include *Il modello del proprietario imprenditore nella Toscana dell'Ottocento: Bettino Ricasoli, il patrimonio, le fattorie*, Florence: 2000.

Salvatore Ciriacono is Professor of History at the University of Padua. His books include *Land Drainage and Irrigation* (Studies in the History of Civil Engineering, Vol 3), 1998.

Juan Díez-Nicolás is founder and president of ASEP, a private consulting firm whose main purpose is social, economic and political research. He is also Professor of Sociology at the Complutense University Madrid.

Brendan Dooley is Professor of History at International University Bremen. His books include *The Social History of Skepticism*, Princeton, 1999, and *Science, Politics and Society*, New York: 2001.

Regina Eich is a researcher at the Forschungszentrum Jülich GmbH in Jülich Germany.

Jürgen-Friedrich Hake is director of the program group 'Systems Analysis and Technology Evaluation' at the Forschungszentrum Jülich. His research interests include issues regarding energy and sustainable development.

Martin Keim is research assistant to the Chair for Macroeconomics at the European Institute for International Economics (EIIW) at the University of Wuppertal.

Dieter Kotte is CEO at Causal Impact, Hamburg, Germany.

Michalis Lianos is Principal Lecturer/CEIST Director at the University of Portsmouth. His current research project is entitled 'Uncertainty and Insecurity in Europe' (EC, European Project, Co-ordinator), CHALLENGE (EC, Integrated Project, Core Partner).

Energy and Culture

Socal Rescoch / Garmany

Petra Lietz is Professor of Quantitative Research Methods at International University Bremen. Her books include *Changes in Reading Comprehension across Cultures and over Time*, 1996.

Power poverty / Estonia

Olev Liik is Dean and Professor at Tallinn University of Technology in Estonia and president of the Estonian Power Generation Society.

Technlgy Research / Macedonia

Natasa Markovska is a research associate at the Research Center for Energy, Informatics and Materials at the Macedonian Academy of Sciences and Arts.

History / US

Martin Melosi is Distinguished University Professor of History, Director of the Institute for Public History, and Director of Graduate Studies in History at the University of Houston. A recent book, *The Sanitary City* (2000) treated the development of water supply, wastewater, and solid waste systems in the United States from colonial times to the present.

Soaslgy / Uk

Linda Miller is employed at the Institute for Employment Studies in the UK. She is currently conducting a review of evidence of the factors that contribute to occupational segregation and will also consider aspects of gender segregation.

/ Macedonia

Jordan Pop-Jordanov is Academician, Professor and Former President of the Macedonian Academy of Sciences and Arts.

Pediatrics / Medicine / Skopje

Nada Pop-Jordanova is Professor of Pediatrics and Head of the Psychophyisiology Department at the Pediatric Clinic, Faculty of Medicine, University of Skopje.

Economics / Germany

Paul Welfens is Jean Monnet professor at the University of Wuppertal and the president of the European Institute for International Economic Relations. His books include *Market-oriented Systemic Transformations in Eastern Europe*, Berlin: 1992; (with G. Yarrow), *Telecommunications and Energy in Systemic Transformation*, Heidelberg and New York, 1997; and (with John J. Addison), *Labour Markets and Social Security*, Heidelberg and New York: 1998.

Meteorology / Bulgaria

Antoaneta Yotova graduated with a degree in Physics, specialization in Meteorology, from Sofia University in Bulgaria. Since 1979, she has been a Research Associate at the Bulgarian National Institute of Meteorology and Hydrology (NIMH) dealing with research in the fields of general and applied climatology, particularly studies on: climate variation and change; greenhouse gas emission estimation; climate and energy interaction, etc. In parallel, she has taken part in a number of national and international projects in her area of expertise, such as the Bulgarian Case Study within the framework of the Inter-Agency Joint Project DECADES (1994-1998) and the UNESCO Encyclopedia of Life Support Systems.

Energy Engineering / Hungary

András Zöld is professor at the Technische Universität Budapest. His research interests include issues of energy efficiency and lifestyle in the mirror of the EDP.

Multi

Inter disciplinary authors, from many diff (European) countries

+ US

Eastern European / West European / US

· Some interest in 'system transformations'

Many not primarily energy engineers - wide range of disciplines are represented.

Acknowledgments

This volume is only the most evident product of a collaboration among individuals and organizations that for the participants has been as pleasureable as it has been fruitful. First of all, we are grateful to the members of the Energy and Culture Research Group at International University Bremen, including Adalbert Wilhelm, Wolfgang Pfaffenberger, Chris Welzel, Petra Lietz, Ryan Richards and Juan Díez Medrano. Over the years since the group's inception in 2003, and especially in recent months, we have shared methodologies, information and points of view, as we carried out a rare experiment in transdisciplinarity. Furthermore we are indebted to IUB, which has generously funded the project, under the leadership of Max Kaase, dean of the school of Humanities and Social Sciences. The March 2004 conference that gave birth to many of the contributions here was funded by grants from the Gerda Henkel Foundation and Bremen Energie Consens, which we thank for their support. Linguistic assistance was afforded by IUB graduate students Erin Palmer and Uros Urosevich. Finally, without the editorial assistance supplied by Kathryn and Joshua Gentzke and Barbara Marti Dooley the work would have been far longer and more difficult. At Ashgate we are indebted to Valerie Rose for her encouragement, and also to Sarah Horsley and Emily Poulton. For the sake of brevity, those many others who have helped we here thank collectively.

My perspective / interest. — T. P. Hughes.

Introduction

Brendan Dooley

Dangers of such a broad canvas.

Energy and *Culture*: the two themes are inseparably linked in human experience, and so they are joined in this book. From the very origins, the production, storage and use of the power to do work have been key features of organized societies. Energy types and the behaviour associated with them, from human to waterpower, from coal and steam to petroleum, have characterized entire epochs. The energy systems on which our societies depended before now have been built by varying mixtures of chance, serendipity, and economic or political opportunism, with little regard for the ultimate costs and benefits to humanity as a whole. In contemporary times these latter questions have become more and more imposing. Energy demand on the upswing due to world demographic growth and the emergence of new economies in Asia and in Eastern Europe raises serious concerns about the ways and means of sustainable development. Projections about the imminent exhaustion of fossil fuels, on which our societies have come to rely, accompanied by the promise of salvation through new fuel technologies, in the midst of increasing fears about environmental degradation, have made energy the key issue of our time. We think this book, by joining a number of literatures that have hitherto remained separate, will make a unique contribution. *AIMS TO CONTRIBUTE TO contemporary debate perhaps* *since 2000,*

Let us begin at the beginning. Anthropologists remind us that human beings as a species rely for survival on the transformation of energy; from food to brains and brawn; from brains and brawn to mechanical acts like turning, drawing, and so on, including every invention the mind has ever conceived, or that people have put into practice, to make cohabitation healthful and convenient. They also remind us that the processes involved in these transformations are multifaceted. Some time after 10000 BC the basic switch from a hunting and gathering to an agricultural type of society appears to have taken place at least in part because of the need to form societies. This is not the place to debate the chicken-or-egg problem of why societies had to be formed in the first place and whether population pressure reduced the viability of one style of energy production (hunting and gathering) and increased that of another (growing crops in a settlement). We will always have to ask, why just then? When and why was a particular demographic density enough to cause change? The same question applies to more recent transitions, say, from wood to coal and from coal to oil; and the same question will apply again to the next transition.

If we enlarge the discourse to consider the behaviour of different peoples in regard to energy transformation, immediately lifestyle considerations come into play. How and when we use energy and in what form depends not only on the supply but also on our imaginations, our desires, and our choices. Let us consider, for the moment, alimentation as an energy source. The historian Vaclav Smil once

Historical perspective on "energy transitions"

[handwritten margin notes at top: "lifestyle", "worldview", "social organisation" } "cultural aspects of energy prodn + use"]

observed the differences in behaviour between two tribes of hunters living in Amazonia in the 1970s. Although one resided in a relatively well-stocked area, and the other in a poorly stocked one, the first ate a poorer diet than the second. The explanation: sheer sloth; and who can argue? Group behaviour varies widely from group to group, as across time and space, often in unpredictable ways. And it may well be true that 'other things being equal, the degree of cultural development [in any given society] varies in direct proportion to the per capita quantity of energy harnessed and consumed' (Smil, 1994, p. 10). However, other things are never equal; and culture itself often seems difficult to apprehend.

[handwritten margin: "Danger of t- Determinism – is recognised"]

In general, what is undeniable is that the quantity of energy harnessed and consumed in a given place and time depends on lifestyle, worldview and social organization. Furthermore, lifestyle, worldview and social organization have determined energy choices long before there existed specific ministries, ministers and policies. The French historian Marc Bloch once showed that the move from animal power to wind power in certain parts of Northern Germany had little to do with the superiority of a new technology. After all, the large property owners who could afford the new technology had no need for cheaper flour. But the opportunity to force all farmers in their regions to grind flour in their mills presented significant and irresistible advantages in terms of political power and prestige. Energy monopolies from then to now have played a similarly ambiguous role. There is no need to embark on overdrawn analogies with modern situations that are only too well known, especially in the American utilities market, in order to invite reflections on the logic of hegemony. A view that only takes the economic dimension into account would be misleading indeed.

[handwritten margin: "Dangers of economic determinism are recognised too"]

The social sciences, in attempting to furnish answers, are challenged not only by the overwhelming complexity of the questions, but also by the wide number of variables affecting energy use and production. In May 2004, specialists in the fields of political science, economics, statistics, history, art history, chemistry and more, originating from Canada, the USA, Bulgaria, Hungary, Germany and elsewhere, converged in Bremen, Germany to discuss energy past, present and future from the widest possible transdisciplinary point of view. In the subsequent months we have had time to compare our results and collect still more information. Combining the insights of the political scientist with those of the economist, the sociologist with the environmental engineer, the historian with the chemical engineer, and keeping in mind the representation of energy issues in the mass media and in modern art, we have asked what are, what have been and what might be the cultural ramifications of energy policy? In other words, how is energy behaviour (production as well as consumption) situated in a human context? Our answers are to be found in these pages.

[handwritten margin: "2004 Conference Bremen"]

We will attempt, we have just said, to situate energy in its human context. We are aware that 'context' may mean different things to different specialists. For economists, it may mean a set of variables in some way related to the chief ones under study. For instance, the chief variable 'efficiency' may be calculated in the light of the related variables 'population density', 'prevalent industries', and 'distribution of energy-saving appliances'. For historians, context may be a set of unrepeatable path-dependent conditions and circumstances, often purely qualitative,

[handwritten margin: "Diff meanings for diff disciplines"]

[handwritten note at bottom: "Focus on cultural ramifications of energy policy / Interest in production + consumption as cultural practices"]

often unquantifiable. In this connection, a tradition of opposition to nuclear power may be as important, for forming attitudes toward energy issues, as a tradition of labour-intensive agriculture.

Our term 'culture' admittedly invites a wide range of definitions. We will adhere to one on which we believe all can agree. Culture is the sphere of experience where values are created and communicated, as distinct from the structures of economy and politics where values operate within organized processes. It involves the transmission of ideas and attitudes through education, mass media and the arts, as well as through everyday practices of gesture, association and conversation. It is composed of the collective memory as transmitted from one generation to the next as well as long-term collective attitudes toward issues and things. Culture, according to our research, is where unexpected outcomes occur; where rational choice occasionally gives way to irrational choice; where the explainable meets the inexplicable. And in the field of energy research, culture, broadly defined, is the reason why predictions often fail. And yet, culture too can be analyzed – as we attempt to do here. By comparing cases and multiplying experiences we have set out to predict the unpredictable, measure the immeasurable. *but inflated*

Culture influences the way people weigh potential benefits and evaluate certain risks, as well as the way they react to the same perceived risks. Differences in risk perception, connected with certain behaviours, appear to be culturally based; and these differences have a strong impact on the position people take regarding the energy problem. Furthermore, cultures of risk in particular places change over time and vary across social groups. There is also a gap between risk as evaluated by experts and as perceived by nonexperts, a gap which in turn influences risk management and its acceptance. Experts' classification of risks according to probability, potential damage, certainty of information, and location of risks, does not coincide with a classification of risks implicit in the perception of the population. Lack of sensitivity to road safety in one place, for instance, may be accompanied by intense concern for other kinds of safety. Similarly, general disinterest in ecological concerns of other sorts may be accompanied by intense concern for the ecological impact of, say, wind power. Even within the same category, contradictions and inconsistencies emerge. A declining perception of nuclear risk, for instance in Italy, may be accompanied by tenacious adherence to a no-nuclear agenda. The globalization of economic and energy developments has not been accompanied by a parallel globalization in the way people assess and act upon energy problems.

From an analytical perspective, the cultural ramifications of energy policy are interesting precisely because of this paradox: in any given time or place, the choice of methods for accomplishing work does not depend on efficiency alone, nor does it depend solely on the probable long-term prospects of the economy – much less, on the future of the biosphere. More complicated causal relations are almost always at work. Shifts in general orientation from one kind of power to another are not directly related to the diffusion of knowledge regarding the advantages of the new source. And where lifestyle and worldview considerations are not decisive, pure political concerns about amassing authority may intervene. Policymakers, in carrying out their responsibilities for managing change, may experience particular difficulty in finding reliable guidelines even in the latter arena, the political arena, where they are directly involved.

Hopefully the remarks so far have sufficed at least to introduce the reader to the multiplicity of concerns in many dimensions that have animated the research reflected in the following pages. Since the subject area is wide-ranging and the kinds of specialized expertise represented are many, we the editors (and contributors) can hardly expect all our readers to be conversant with the key contributions in each separate subsection of the book, even if they are (as they probably are not) dedicated followers of journals like *Energy Policy*, *Energy*, *Journal of Energy Development*, *Energy Conversion and Management*, *Urban Affairs Quarterly*, or even (with particular regard to significant technological developments) *Nature* or *Science.* Although almost everyone has probably heard of Kenneth Deffeyes, *Hubbert's Peak*, and perhaps even read with passion Lester Brown's *Eco-Economy* (Norton 2001), those who are not sociologists might not be fully informed about Mary Douglas and Aron Wildavsky's *Risk and Culture* (Berkeley 1982), even if, as economists, they are well aware of Carlo C. Jaeger, Ortwin Renn, Eugene A. Rosa, Thomas Webler, *Risk, Uncertainty and Rational Action* (Earthscan 2001), or, as statisticians, with Benjamin F. Hobbs and Peter Maier, *Energy Decisions and the Environment: A Guide to the Use of Multicriteria Methods* (Kluwer 2000). Nor will non-historians and non-sociologists necessarily have David E. Nye, *Consuming Power: Social History of American Energies* (MIT 1997) or Gabrielle Hecht, *The Radiance of France: Nuclear Power and National Identity* (MIT 1998) directly at their fingertips, even if their interest in environmental concerns has led them in the direction of now-classic works on environmental history like Carolyn Merchant's recently republished *The Death of Nature*.

The more we have discussed and compared notes, the more we realize how much still needs to be done. For a synthetic picture of future projections, the interested reader might refer to the one traced by Vincent Lauerman, *Energy Scenarios for the 21st Century* (CERI 2002), and similar studies; but those seeking specific data on current usage and production patterns, in the current absence of significant up-to-date international studies on the subject, will have to resort to specific national industry and company surveys as well as national associations such as the Association of Bulgarian Energy Agencies, the French Réseau des Agences Régionales de l'Energie et de l'Environnement, the Association of Spanish Energy Management Agencies, the American Environmental Protection Agency and Department of Energy. Concerning consumer behaviour and lifestyle, the Euromonitor Global Markets database is indispensable, although not sufficiently nuanced with respect to regions and cities. Concerning public opinion regarding environmental and energy issues, the World Value Survey and Euromonitor remain the platforms for all further study.

Many of our readers will share our feeling of indebtedness to studies by Anthony Giddens, Manuel Castells, Ulrich Beck and others, regarding the structural origins of current attitudes to modernization, technology, the environment and post-materialist values. We now know that a major attitude change has occurred across Europe and America in the last decades, in response to globalization of markets, the acceleration of scientific and cultural change, and the

Need for more cultural analysis of energy issues (handwritten)

application of science and technology to an ever wider range of areas affecting citizens. Perhaps readers will share our view that much more needs to be done to analyze the social basis of opinion, taking into account variations from place to place and across social groups and subcultures. Only in this way can the cultural roots of national differences on these issues. We need more empirical studies perhaps along the line of Ronald Inglehart's work; and we need local studies on value production past and present, like Robert Putnam's, but taking energy and not just politics into account. We know as well as many of our readers that there are many specific features of context, still elusive but essential for understanding and predicting variations in outcomes, that could have a powerful impact on energy policy.

Some recent perspectives are inspired by a radical critique of state and economic development in the West. On the one hand, drawing upon the poststructuralist view of power, they see the state not as a given, but as the particular product of a line of development that has characterized Western politics. State institutions, they argue, impose artificial hierarchies not only on the humans subject to them but on the entire planet that is under state control. The new focus on environmental concerns, instead, calls for thinking of humans and their communities in a wider pluralist context of harmony with nature. On the other hand, drawing on a post-capitalist view of economic development, they consider modern industry and the economies stemming from this as the product of the privatization of our planet's natural resources. A more community-minded outlook, they suggest, would take account of the common interest of all humankind in resources and environment, and to this ought to be added the interest of all in a more equitable distribution of prosperity. We will see occasional reflections of these ways of thinking in the following pages. *Relevance for today?* (handwritten)

The reader at this point will probably be wondering exactly how the varieties of culture vis a vis energy issues may help us understand current problems. And at least for the issue of environmental legislation, and the specific portion of culture related to political and economic structure, there is a useful study by Lyle Scruggs (2003) entitled *Sustaining Abundance*. This multivariate analysis of seventeen countries on three continents considers the impact of various socio-political conditions in bringing about specific outcomes in terms of legislation on greenhouse gas emissions, recycling, waste disposal, fertilizer and the like. And according to these results, societies with more centralized democratic institutions, and societies where economic interest groups operate in close cooperation with governments (the 'neocorporatist' model), tend to perform better at the policy level, where environmental issues are concerned. Still to be determined is the role of culture as a whole (education, representation, history, and so forth) in bringing about desired outcomes; and especially, we want to know what is the role of these features in bringing about policy changes where issues specifically related to energy production and consumption are concerned. *Adv of neo-corporatism* (handwritten margin note)

For understanding these latter aspects, i.e., the cultural background of energy issues, and the role this may play in energy policy, we must defer to our collaborators. Each of them has taken a different aspect of the energy and culture complex and defined it in his or her own way. The aspects we have chosen – politics, history, lifestyle, risk, representation and science – hopefully cover a good

book sections (handwritten)

portion of the modern debate. The book is therefore organized systematically according to the various treatments of these aspects, including, within each section, analyses utilizing different modes of research, different data and different geographical and chronological contexts. In some cases the divisions may seem slightly artificial; and there is naturally a considerable amount of overlap between one section and another. After all, risk is as much an issue for scientists, our contributors have found, as it is for economists; lifestyle questions are important aspects of the political scientist's apprehension of energy issues, just as they are for the historian; scientists, educators and historians are concerned with the political and human resources implications of energy problems; and all fields are affected by the way in which energy is represented in the mass media and in visual communication. On the whole, we view our endeavour as a multi-angle perspective on a single great problem.

We thus begin with the issue of Energy and History. Indeed, use of energy resources from the outset of civilization has depended to a large extent on social, cultural and political factors; and the disappearance of one resource has necessitated the development of others, always involving large-scale readjustment and negotiation with unforeseen consequences. Martin Melosi considers the concept of energy transitions as a category for understanding energy cultures past, present and future. The supposed 'advantages' of one energy source over another, given the demographics of the time, have never been the sole cause of change; nor has any period been characterized entirely by a single source – instead, in each period some combination of older and newer sources usually came into play. How did change take place? Melosi shows how marketing techniques and lobbying contributed to the emergence of oil power, while nuclear power came about through the government military program. In the transition to the energy configuration of the future, he notes, non-economic incentives will be highly important. Salvatore Ciriacono agrees, adding that the latter stages of a passing energy regime may be accompanied by economic stagnation and even social disruptions, unless a new regime emerges in a timely fashion accompanied by the technical and economic conditions for its development. Through the centuries, he reminds us, efficient use of energy has been a key to economic growth; and the decisive factor differentiating dynamic from depressed areas. European economic superiority, he argues, was built upon the availability and advanced use of water power; and a closer examination of this often-overlooked form of energy sheds new light on one of the most important transitions in the Industrial Revolution: from water to steam – a transition which according to Ciriacono's evidence was far more gradual and complicated than is commonly perceived. Giuliana Biagioli reminds us that recent generations are not uniquely responsible for environmental degradation; it was noticed already at the beginning of the Industrial Revolution. That was when a thousand-year-old production system in the Mediterranean basin, at once parsimonious and ecological, began to give way to more productive, yet also more energy-intensive and wasteful mechanized processes. Meanwhile, even where these processes did not become widespread, such as in the typical hilly areas where labor-intensive methods continued to predominate, nonetheless, ancient cultivation and conservation techniques gradually died out with nothing to replace them.

Access to sources of energy has often been mediated by politics, as our contributors note in the section on 'Energy and Politics'. In modern societies, energy policy requires continuous negotiation involving industries, labor organizations, non-governmental organizations, and the general public. Jürgen-Friedrich Hake and Regina Eich offer a broad overview of the concept of 'sustainable development' and its status in the literature since the so-called Brundtland Commission in 1987. Although agreement upon a valid set of indicators has proven particularly difficult, nonetheless there is substantial agreement about the 'necessary limitation of anthropogenic material inputs into existing ecological cycles'. Environmental damage, in a globalized world, is measurable by costs to the economy. Nor should this issue entirely crowd out the equally important issue of maintaining secure energy supplies for the future, as has so often happened in recent debates for instance in Germany. Bernard Beaudreau illustrates the heuristic value of a theory of political economy with energy at the center for understanding current problems. Energy incomes, he suggests, constitute the very basis of material wealth, and so they directly affect economic growth. In recent times, the reduction in energy incomes in highly industrialized countries has either caused or contributed to recent events like the productivity slowdown, increasing income inequality and increased social and political tensions. Negotiation between the various interests is now carried out for shares of a less and less rapidly expanding national wealth. Efforts to compensate for the reduction in energy incomes have helped bring about a questionable (at least from the productivity standpoint) ICT revolution, globalization and the relocation of production to low-wage countries, exacerbating the other problems. Paul Welfens and Martin Keim examine the European Union electricity market liberalization and regulatory policies in the light of the different political, economic and industry structures in play. In spite of good intentions, they suggest, real liberalization is likely to encounter significant obstacles. On the one hand, politicians' environmental agendas are likely to get in the way within each of the countries. Moreover, liberalization to be effective in reducing prices has to be enforced; and there are signs, for instance in Germany, that a limited number of large long-standing firms will retain control.

The scientific imagination from the outset of modern science has been captivated by the promise of cheap, efficient power. In modern times, meeting the mandates of energy consumption and environmental protection has encouraged innovations in the areas of physics and chemistry for the realization of suitable alternative technologies. However, developing new technologies for renewable energy requires qualified scientists and engineers. Yet scientists and engineers represent a relatively small proportion of the workforce in the EU. Petra Lietz and Dieter Kotte begin the section on 'Energy and Science' by asking the question, why should this be? They suggest that the different quotients of science careers across the European countries indicates a context-based explanation. Indeed, their data demonstrates interesting differences from country to country regarding the relation between students' initial expectations of a career, students' aptitudes, students' backgrounds and actual job placement. For instance, where a significant gap exists between desired and actual engineering and science careers, especially in low GDP countries, career motivation may be more related to economic self-improvement than to perception of real ability. In the authors' view, a better knowledge of these characteristics may help countries and officials develop policy solutions, and to form educational and public awareness programs capable of attracting more students to careers in these fields. Linda Miller focuses on the science career gap between males and females, pointing out that increasing the participation of women in these careers would significantly improve the total number of scientists and engineers. Indeed, young girls tend to outperform boys even in the sciences, although they seem less inclined to pursue careers in these fields. Family dynamics obviously plays a role. In addition, data show a correlation between knowledge about a particular career and a penchant to pursue it. Thus, the often inaccurate and misleading view of science put across in the mass media and even in the classroom may be a deterrent. She suggests more research into teaching styles and course content on one hand and, on the other, better career guidance.

If new energy sources are to be effective they must be adopted and applied by people; and environmental protection begins with the individual citizen. Thus, 'Energy and Lifestyle' are closely linked, also in the contributions to the section bearing this name. Indeed, patterns of energy use depend on behaviors that are as much rooted in geographical and cultural as in chronological contexts. In Bulgaria, Antoaneta Yotova notes, programs have been developed to increase the relatively minor role played by renewables, but there are significant obstacles. Electricity is already high-priced; and investment is costly. Political engagement to carry out the more environment-friendly provisions of the Law of Energy (2003) is weak, and there is little information or encouragement among the general public to incite officials to more decisive action. Only the country's planned entry into the EU and the obligatory adherence to the Kyoto protocol and the EU targets are likely to bring about change. In Estonia, Olev Liik shows, the chief energy asset is also the chief limitation: namely, the presence of considerable oil shale deposits. These deposits, which account for some 65 percent of the primary energy supply of the country, have made Estonia energy independent; at the same time their easy availability has discouraged the development of renewables. Yet Estonia's participation in the 2000 Genf convention concerning long-range transborder air pollution, along with the EU directive of 2001, mandate significant reductions in sulphur dioxide emissions by 2005; so the use of sulphur-rich oil shale will have to be severely curtailed. András Zöld directs attention to an issue that is currently preoccupying building engineers and architects, and not only in Hungary, where he writes – namely, the construction and updating of energy sustainable structures in conformity with the EU's Energy Performance Directive of 2002. By January 2006, he reminds us, member states are obliged to devise and implement legislation in conformity with this directive. For dealing with the practical problems, he suggests a combined approach, taking account of different types of buildings, different systems of climate control, and different users.

From the emergence of hydroelectric power to the building of nuclear reactors, new sources of energy have often raised issues concerning the risks to populations and the environment. Culture influences the way people evaluate certain risks, the way they weigh potential benefits and risks associated, and the way they react to the same perceived risks. Furthermore, cultures of risk in particular places change over time and vary across social groups, and between expert cultures and the population at large. Opening the section on 'Energy and Risk', Michalis Lianos points out what he sees as the well-advanced 'dangerization' of our life-world. The sensation of uncertainty derives as much from actual risks as from perceived ones. Very real risks to our planet's survival and our own have induced the sensation of uncertainty that permanently accompanies the 21st-century consciousness. This situation, Lianos suggests, should encourage us to adopt the 'environmental horizon', taking into account humanity's responsibilities as a standard in political, social and economic interaction. Natasa Markovska, Nada Pop-Jordanova, and Jordan Pop-Jordanov, on the other hand, note the particularly high psychological impact of energy development in Eastern Europe. Especially among employees involved in the energy industries, the effect of information overload concerning risk factors can lead to stress-related disorders. The authors report on the success of a regime of biofeedback training for alleviating these symptoms.

Opinion regarding energy issues is an important element both of mass culture and of elite culture. The mass media not only heighten public awareness of energy and environment issues, but spark public debate. However, as Juan Díez-Nicolás reminds us in the section on 'Energy and Opinion', ideas about the issues are unevenly distributed within society, and there is no direct connection between, say, possessing environmental awareness and carrying out actions to protect the environment or even expressing this awareness in some active way. Utilizing survey data compiled in Spain, he shows, people's convictions about the importance of saving energy and protecting the environment may be more or less superficial; in any case, more often what governs their actual conduct are the immediate concerns of family economy.

A number of themes keep reemerging in these pages, as seen from different points of view. For instance, the notion of prevailing energy regimes comes up again and again. Ciriacono and Melosi place this in a historical perspective, noting that prevailing energy systems have always relied on a complex mix of different energy sources, not one source alone; and the mix may include some combination of old and new technologies. But the most important agreement concerns economic development. And it would be difficult to overemphasize the conclusion reached by all participants, that energy development and economic development work absolutely in tandem. Beaudreau in fact goes so far as to place energy development at the center of his new theory of growth.

Not all collaborators agree. Michalis Lianos questions the way Western society has embraced 'power' as its telos in every sense: from the industrial to the political to the international. Others see power as the key to economic development, and development as the essence of progress. Likewise, where Lianos emphasizes the real risks of environmental damage, Markovska/Pop-Jordanov emphasize the 'psychological pollution' caused by exaggerated perceptions of risk induced by prevailing environmental discourse. Even when collaborators agree in principle, they may disagree in emphasis or in the different weight to be attached to different aspects of the general topic. Beaudreau notes that historically, private control over natural energy resources has been the cause of the great accumulations and concentrations of industrial capital. Biagioli agrees that economic development and energy development have gone hand in hand; but she focuses more on what is lost than on what is gained. At least in the countryside, modern productive practices have taken away more than they have replaced, and technology still has provided no substitute for older systems that utilized the land while preserving it.

The coming into force of the Kyoto accords in February 2005 has imparted a new urgency to discussions about Europe's energy future. Now more than ever, European governments are challenged to help define the kind of society we are, and the kind of society we want to become. However, policymakers and the general public need not make decisions and recommendations on the basis of misleading and incomplete information or analysis, merely because the verification of such data takes place within the restricted precincts of expert cultures. Quite the contrary. The complexity of the problem and the urgency of the result demand a more fruitful partnership between knowledge specialists and the public. The social sciences and humanities are in a unique position to inform the debate; and we hope this book will contribute to the discussion at least as much as its compilation has inspired its authors to deepen their transdisciplinary expertise.

References

Beck, U. (1992), *Risk Society: Towards a New Modernity*, Sage Publications, London.

Bloch, M. (1939-40), *La société féodale*, 2 vols., A. Michel, Paris.

Brown, L. (2001), *Eco-Economy: Building and Economy for the Earth*, Earthscan, London, UK.

Castells, M. (2000), *The Rise of the Network Society*, 2nd ed., Oxford University Press, Oxford, UK.

Deffeyes, K. (2001), *Hubbert's Peak,* Princeton University Press, Princeton, NJ.

Douglas, M. and Wildavsky, A. (1982), *Risk and Culture*, University of California Press, Berkeley, CA.

Giddens, A. (1991), *Modernity and Self-Identity: Self and Society in the Late Modern Age*, Stanford University Press, Standford, CA.

Hecht, G. (1998), *The Radiance of France: Nuclear Power and National Identity*, MIT Press, Cambridge, MA.

Hobbs, B.F. and Maier, P. (2000), *Energy Decisions and the Environment: A Guide to the Use of Multicriteria Methods,* Kluwer Academic Publishers, Dordrecht, Netherlands.

Jaeger, C.C., Renn, O., Rosa, E.A. and Webler, T. (2001), *Risk, Uncertainty and Rational Action*, Earthscan , London, UK.

Lauerman, V. (2002), *Energy Scenarios for the 21st Century,* Canadian Energy Research Institute, Calgary, Canada.

Nye, D.E. (1997), *Consuming Power: Social History of American Energies*, MIT Press, Cambridge, MA.

Scruggs, L. (2003), *Sustaining Abundance: Environmental Performance in Industrial Democracies*, Cambridge University Press, Cambridge, MA.

Smil, V. (1994), *Energy in World History*, Westview Press, Boulder, CO.

PART 1:
ENERGY AND HISTORY

Chapter 1

Energy Transitions in Historical Perspective

Martin Melosi

The energy historiography of the United States has been strongly influenced by the concept of 'energy transitions'. So far, the concept has helped to periodize the development and use of energy sources over the years, and to explain economic and environmental changes taking place, particularly since the Industrial Revolution. More recently, the concept has been adopted in discussion about future energy development and use, with an eye toward promoting sustainable development. This chapter will discuss various uses of the concept and its promise for clarifying current issues.

The concept of 'energy transitions' is based on the notion that a single energy source, or group of related sources, dominated the market during a particular period or era, eventually to be challenged and then replaced by another major source or sources. By tracing the rise and fall of dominant types of energy, scholars have focused more on societal shifts than on the inherent characteristics of rising and falling energy-source dependence.

As an historical tool, the energy transition has much to recommend it. In the broadest sense, the concept can help researchers understand the evolution of human material culture, economic growth and development, the utilization of resources, and social organization. Utilized too narrowly however, it may merely provide a convenient instrument for segmenting energy history within a one-dimensional chronology. The most beneficial way to utilize the conceptual tool of 'energy transition' would perhaps be to see this concept as a study of a fluid process, rather than understanding it as a pretext for establishing rigid barriers between specific energy eras. Potentially, energy transitions can help to clarify how energy development and use influences and is influenced by the technical, economic, political, environmental, and social forces that shape society.

Energy transitions can be used as the basis for historical arguments in two ways: (1) quantitatively – by measuring changes in energy consumption, or (2) qualitatively – by evaluating the impact of new sources of energy on various aspects of American life. The quantitative approach yields a simple periodization of energy history based on peaks and troughs of wood, water power, coal, petroleum, natural gas, and so forth. By illustration, Table 1.1 shows some basic aggregate statistics for the use of wood, coal, and oil in the United States from 1850 to 1955. This widely used model traces the energy history of the United

States through cycles of supply and demand with particular attention to increases in energy consumption. It recognizes two major energy transitions during the period: from wood to coal in the nineteenth century, and from coal to petroleum in the early twentieth century. (Many other simple statistical studies focus on energy production.)

Table 1.1   Percentages of aggregate energy consumption for wood, coal, and oil in the US

Year	Wood	Coal	Oil
1850	90.7	9.3	0.0
1885	47.5	50.3	0.7
1910	10.7	76.8	6.1
1930	6.1	57.5	23.8
1955	2.6	28.7	40.0

Source: Sam H. Schurr and Bruce C. Netschert, *Energy in the American Economy, 1850-1975* (Baltimore, 1960): 36

In the first energy period, the use of wood (and to a lesser extent wind and water power) is seen to dominate the United States until the mid-1800s, with wood consumption peaking in 1885. Beginning in 1850, the use of coal was established on a commercial basis, and by 1885 coal mining became a major industry. Coal, in terms of national consumption, dominated the energy market from the mid-1880s through World War I. At that point coal began to suffer a relative decline, while petroleum (and later natural gas) became a more popular source that undermined coal's dominance. Petroleum usage moved through several stages of its own before supplanting coal (to refine the quantitative approach even further). In its 'kerosene period' in the years after 1869, it became a major illuminant; in the 1890s it passed into its 'fuel oil period'; by the turn of the century it became an important internal combustion engine fuel (Schurr and Netschert, 1960).

This periodization emphasizes broad patterns of extraction and consumption and has furnished a background for economic analysis of contemporary and future energy requirements. Many scholars have accepted the quantitative approach to energy transitions uncritically and with little consideration for the causes and effects of those transitions in general. However, this approach alone does not sufficiently explain why and how energy choices are made. Factors such as shifts in availability, technological advances, changes in the nation's output of goods and services, and shifts in consumer preference are not adequately explored but subordinated to the period of the transitions (Schurr and Netschert, 1960).

The qualitative approach – also relying on statistical data – offers greater opportunities for understanding the impact of energy sources on society. Studies have drawn distinctions between the 'traditional' energy eras. For example, several writers emphasized that energy use in the United States shifted from 'renewable' (wood, wind power, water power) fuels to fossil or 'non-renewable' (coal, petroleum, natural gas) fuels. This focus provides a good perspective for evaluating the impacts of energy transitions, namely resource depletion, environmental impact, changes in habits of consumption, and alterations in governmental policy, but the terms need to be clarified. The concept of a 'renewable' resource, for example, cannot be taken at face value, especially until a variety of 'costs' entailed in developing and utilizing it are understood. The renewable/non-renewable perspective has been used only in a cursory manner in dealing with historical trends, and has tended to enter the discussion in the wake of the energy crisis of the 1970s and the modern environmental movement ('energy strategies', 'energy futures', 'alternative energy paths' and 'energy forecasting'). (Perelman, Giebelhaus and Yokell, 1981; Stobaugh and Yergin, 1979; Schurr et al., 1979; Energy Policy Project, Ford Foundation, 1974; Kranzberg, Hall and Scheiber, 1980; Goodwin, 1981.) Future projectors have made some thought-provoking speculations, for instance that a long-term transition to a solar-based system could produce ominous societal consequences – possibly a return to feudalism and energy sources based on territoriality (Perelman et al., 1981). This kind of speculation, however, while relying on some historical trends, deals only tangentially with the significance of past transitions.

Although the qualitative approach to energy transitions has been primarily a response to the 'energy crisis' and beyond, its emphasis on energy transition as a process has attracted some attention from historians. One useful version examines how economic factors, which influenced the transition from one source of energy to another, are conditioned by non-economic factors in producing change (Nye, 1998). For example, some studies of the rise of petroleum in the United States in the late-nineteenth and early-twentieth centuries argued that oil not only had a competitive edge over coal in terms of lower price, but also in terms of the structure of the industry. Oil companies developed superior technology for exploiting oil production and developed effective marketing techniques and more reliable transportation (rails and pipelines), and were able to promote their product as a more versatile fuel than coal. Oil's economic advantage was tied not so much to the development of national markets as to the rapid acquisition of new markets, especially in the emerging 'sunbelt' region of the South and West (Pratt in Perelman et al.; Tarr and Lamperes, 1981). This type of analysis expands the use of energy transitions by examining the *mechanism for change.* It also raises questions about the nature of fuels as competing, rather than as complementary, energy sources. Similar studies have explored the transition from wood and water power to coal, focusing on the development of a wood-based society and the difficulty of coal or other fuels in creating an economic, social, and political infrastructure to challenge the existing system.

Another fruitful approach is to emphasize the evolutionary as opposed to the revolutionary aspects of the energy transition. Specifically, older sources of energy (muscle power, renewable resources) are not replaced totally by newer sources (fossil fuels, nuclear energy). Instead, they are supplemented, complemented, or slowly displaced according to use. Calculating the total consumption of an energy source gives little sense of the stages of transition, and creates artificial thresholds through which society apparently passes from one energy era to another. Energy eras, therefore, might best be viewed as the accumulation of many smaller transitions.

aggregate tendency of quantitative
transitions approach disguishes/veils
many of the multiple realities of transit

The First Energy Transition

America's first major energy transition, which took place in the nineteenth century, is an excellent example of this evolutionary process. Over the course of several decades, muscle power and renewable fuels continued to be employed for several tasks despite the rise in the use of anthracite and bituminous coal as industrial, and then domestic, sources of energy (Greenberg in Frese and Judd, 1980; Cochran, 1981). The factors that helped to perpetuate the use of older sources and the experience of newer ones provide some useful avenues of inquiry about the nature of transitions.

perpetuity older patterns while new ones emerge

The authors of *Energy in the United States* (1968) offer a simple explanation for the first major energy transition: 'It was not until the continuing cutting down of forests had raised the price of wood and removed its sources further and further from the emerging centers of population and industry that coal came to be accepted more widely … What gave it the crucial push was the use of bituminous coal for two rising segments of the modern economy-steel manufacture and steam generation' (Landsberg and Schurr, 1968). Although widely accepted at one time, this view depends too heavily on national patterns of consumption and fails to take into account the more complex nature of energy transitions. America's initial energy transition spanning the nineteenth century was clearly more evolutionary than revolutionary because: (1) in the early- to mid-nineteenth century the United States was dominated by local, regional, and sectional interests making the transition dependent on local availability, use, and preference; (2) the transition was from wood, water power, and wind power to anthracite and bituminous coal, i.e., from an array of 'renewable' resources to two quite unique fossil fuels located in different sections of the country and possessing different properties as fuels; (3) variables other than the depletion of wood resources led to the transition, including available transportation, market factors, technological changes, and head-on competition between the various types of fuels. The emergence of a new form of transportation or a new industrial process, however, did not necessarily mean a change in energy source; the question of adaptability influenced an energy source's continued use.

macro or micro

more complex growth process

interplay of technology

What led to the increasing use of fossil fuels in the United States was the transformation of the country from a rural, agrarian, decentralized society into an urban, industrial, national culture. Coal did not bring about that transformation, but in the long run it adapted more successfully to it than the renewable sources of energy, which were much more territorially bound and less versatile. While renewable sources retained a portion of the industrial and domestic energy market, an increasingly urban-based industrial economy, a rapidly expanding national

chicken + egg

railroad network, and a highly concentrated population came to depend on coal (Hindle, 1975; Martin, 1976; Taylor, 1951; Lillard, 1947; Cole, 1970; Petulla, 1977; Nettels, 1962, Burlingsam, 1938; Hunter, 1949; Tillman, 1978).

The first American energy transition demonstrates the unrealized potential of the energy transition as a historical tool. The nature of nineteenth-century American society made it impossible for a new energy source to revolutionize industry, transportation, and home heating overnight. The series of smaller, albeit significant, changes in patterns of use occurred for many reasons, including consumer preference, availability, relative cost, technical innovation, and geographic determinants. In a heterogeneous nation that spanned a continent, a single, convulsive energy transition was impossible. Yet while the causes of the transition produced evolutionary change, the results were indeed revolutionary.

[handwritten margin note: evolutionary & multi-factoral causes, but revolutionary, transformative results.]

A Third Transition?

Space does not permit us to develop the background behind the first energy transition, nor to give detailed attention to a second transition. Let us now turn instead to more recent times. Has a third transition already occurred, and if it has not, is one imminent? To what degree can historical analysis reveal new patterns in energy development and use? In the 1980s, in some early inquiries into the notion of energy transitions, it was asserted that the United States had entered a third transition from petroleum and natural gas to a wide array of alternative fuels in the late 1960s. Despite the persistence of a petroleum-based culture, it was argued, the control of oil by OPEC and other powers forced Americans to explore the viability of alternative energy sources. Unlike the first two transitions, the most recent one, it was added, had less to do with a shift from one major energy source to another, and more to do with a changing perception of the role of energy in the life of the nation (shaped by energy use and development patterns in the 1960s), the 'energy crisis' in the 1970s, and the emergence of the modern environmental movement. This assessment assumes that the 'energy crisis' was a factual reality rather than a mere perception, that the forces controlling petroleum and natural gas would continue to squeeze consumers, and that projections about the availability of petroleum in particular indicated a major shortfall within ten years or so. These projections proved to be inaccurate. In fact, the 'energy crisis' was not a crisis in long-term supply. Forces controlling energy sources were as dependent on their markets as they were on control of supplies. And projections about available petroleum and natural gas supplies in the future have changed several times.

[handwritten margin note: use (& mis-use) of history]

[handwritten margin note: 1970s energy crisis as a mere "perception"]

Since that time, dependence upon petroleum and natural gas has not abated, but an older mindset has been challenged. This older mindset, powerful in the 1960s, fostered a belief in unbridled material progress linked to visions of a high-technology society, where every family drove a new car, had a modern ranch-style home, and accumulated every appliance that could be plugged into a wall socket. Since then, the rise of the modern environmental movement and a variety of economic dislocations – the energy crisis among them – has challenged the uncritical notion of an optimistic materialistic future. The rethinking of energy

[handwritten margin note: Mindset change]

[handwritten margin note: rise of environmental movement]

development and energy needs – often among policy makers, interest groups and reformers, but rarely the general public – did not constitute an energy transition per se, but indicated the kinds of issues that were likely to influence one at some point in the future. Two examples are suggestive: (1) the rise and fall of nuclear power in the United States (the expectation of a limitless source of power, but one fraught with risks); and (2) the idea of a 'hydrogen society' replacing conventional fossil fuels (an outgrowth of the rising dialogue over sustainable development).

The promotion of nuclear power in the 1940s and 1950s was not linked to any perceived need for additional sources of power. Instead, governments, inspired by political and diplomatic interests, pressured a reluctant and unenthusiastic private sector to develop commercial uses. Eisenhower's 'Atoms-for-Peace' speech is significant. On December 8, 1953, the President appeared before the UN General Assembly and proposed that the nations capable of producing fissionable material contribute to a pool from which other nations could draw for nonmilitary purposes. This meant sharing civilian nuclear information. The speech also called for an International Atomic Energy Agency (IAEA). Some observers viewed the plan as a way to create a large enough demand for fissionable material to limit what otherwise might be used to build weapons. Taking the argument a step further, others argued that Eisenhower while pointing out the horrors of nuclear war, was trying to offer hope through expansion of the civilian uses of atomic energy. The appeal, through indirect means, sought some form of arms control and possibly disarmament. On a more pragmatic level, Atoms-for-Peace would allow the United States to maintain its commanding lead over the Russians, neutralize potential adversaries, and also provide a great propaganda outlet (Melosi, 1985).

By taking the lead in developing peaceful uses for the atom, the United States could retain its overall leadership in nuclear power. While Atoms-for-Peace was good propaganda, the United States followed through with only mild support for the IAEA. When cheap oil became available from the Middle East, the idea of nonmilitary nuclear power development on a world scale petered out.

Given the events of the 1950s, it is amazing that commercialization of nuclear power occurred at all. But with the passage of the Atomic Energy Act of 1954, nonmilitary applications received a substantial boost. The energy market had little to do with this important event, since there was no pressing need for a new source of power in the United States. There was, however, strong interest in enhancing American prestige. Several government leaders believed that the United States ought to lead the way in the development of nuclear power, especially beyond the making of weapons. Since nuclear power was a government-dominated and government-sponsored energy source, the mechanisms for carrying out that goal – the Atomic Energy Commission (AEC) and the Joint Committee on Atomic Energy (JCAE) – were already in place (ibid.).

The 1954 Atomic Energy Act combined the spirit of Atoms-for-Peace with a call to develop civilian nuclear power in the private sector. It authorized greater international cooperation for the AEC, by providing for more latitude in the dissemination of scientific data. It also sought to increase participation by private enterprise in the development and construction of reactors. Private firms would be allowed to own reactor facilities, with the government retaining ownership of the nuclear fuel (ibid.).

Those countries that have had the most successful nuclear programs have been those isolating NP from the market most effectively

Energy Transitions in Historical Perspective 9

The AEC was not retreating from nuclear power development altogether. It retained most of its regulatory powers, for instance, issuing permits for entering the industry and constructing facilities, maintaining control over the industry's nuclear products outright by purchase and by issuing permits to purchase and or setting security and safety standards. In 1962, with among other things, the indemnification of private companies having been secured in 1957, the AEC issued a report claiming that nuclear power was now commercially viable. But the announcement created no waves of enthusiasm. Indeed, the report was more a hope than a reality. In 1961 two-thirds of the reactor programs still emphasized weapons and other military applications. Nonetheless, the AEC and congressional supporters of nuclear power continued to promote commercialization and began achieving results. Despite the fledgling industry's errors in estimating capital costs, demand for nuclear power facilities was on the rise in the mid-1960s. High coal prices and growing sensitivity to pollution associated with the burning of coal made nuclear power more attractive to utility companies. The Northeast blackout of 1965 also stimulated interest in nuclear power generation. By the end of 1967, American utility companies planned to construct seventy-five nuclear plants; about half of all power plant capacity ordered was nuclear. By the end of 1969, ninety-seven nuclear plants were in operation, under construction, or had been contracted. For the moment, the future of the young industry looked bright (Del Sesto, 1979; Aviel, 1982; Perry, 1977; Allen, 1977; Dawson, 1976; Mazuzan, 1981; Keating, 1979; Gillette, 1972). *bright prospects in 1960s*

Although national energy policy showed few signs of change in the 1960s, the environmental implications of energy development and use acquired greater significance. The increased utilization of coal, oil, and nuclear fuels, due to the scale of electric-power production and the rising demand for energy, drew attention to the exploitation of natural resources. The commercial viability of nuclear power raised questions about radiation, plant siting, and reactor safety. And the ubiquity of air pollution – especially from automobile emissions and stationary fossil-fuel burners – moved policy makers toward more effective clean air standards. Although these issues were treated as separate and unique problems for most of the sixties, they were significant enough in scope to bring the relationship between energy and environment to national attention (Melosi, 1985; Andrews, 1980; Petulla, 1980).

Another environmental crisis arose in the late 1960s as opponents of nuclear power seized on its health and safety risks. Nuclear safety had been a secondary issue in the early development of nuclear power. The race to produce the bomb, the quest for strategic superiority engendered by the Cold War, and the determination to commercialize atomic energy received top priority. Nuclear power safety grew as a public issue in the 1960s, especially after a small test reactor exploded at the National Reactor Testing Station in Idaho Falls in January 1961. The AEC claimed that the resulting three fatalities were due to an electrical power-surge blast; union officials blamed them on radiation. Nonetheless, the AEC established guidelines for siting nuclear reactors away from large urban populations and in 1962 set up a procedure for relating plant size to distance from dense populations (the concept of 'remote location'). Late in the decade, however, the remote location concept had still not evolved into a clear set of standards. Taken as a group, the energy and environmental issues of the 1960s offered serious challenges to the American people and their government. But public acknowledgement of the rise of the modern environmental movement and the advent of the energy crisis in the 1970s provided greater focus.

The conflict over nuclear power in the 1970s developed on two levels: one centered on the nature of the energy source itself, the other on the role of centralized power. Debate on the first level revolved around the question of safety. On the second level, broader societal and institutional issues were at stake. Advocates touted nuclear power as an answer to the energy crisis and to OPEC's control of oil, characterizing dependence on coal as a greater environmental risk than nuclear power. But antinuclear groups, suspicious of centralized power production through large nuclear systems, argued that such production kept energy development in the hands of the government and big business and left consumers vulnerable to their whims. They called for decentralized systems – especially solar energy – which they argued would not only reduce the need for nuclear power, but also weaken the trend toward corporate control of society.

Although the war of words grew more intense, environmentalists scored some important victories for the antinuclear cause with protests over several sitings of nuclear power plants. The safety question also remained the most controversial. The AEC decision to push ahead with the breeder reactor raised serious questions about the production of highly toxic plutonium – a requisite ingredient in atomic weapons. Through the early 1970s the AEC had treated the safety program as an in-house matter, but this became more difficult as environmentalists demanded a public accounting. Ultimately, the public furor over safety forced the AEC onto the defensive, leading to the Reactor Safety Study – or the Rasmussen Report (directed by MIT nuclear engineer Dr. Norman C. Rasmussen) in 1972. The study was largely an apologia for nuclear power, concluding that the risk from nuclear reactors was very small, and that chances of core meltdowns were particularly remote (1 in 20,000). Contrary to AEC hopes, the Rasmussen Report raised more questions than it answered, but for nuclear advocates the lack of any catastrophic accident was vindication or at least recognition that criticism of the safety program was unduly alarmist (Barlogh, 1991).

If the Rasmussen Report proved useful in defending the safety programs of the AEC, the onset of the energy crisis offered the chance to promote nuclear power as a hedge against OPEC and the scarcity of oil. By late 1973 the five-fold increase in imported oil prices made nuclear power competitive again; one year later orders for light-water reactors reached a new peak. But almost as quickly as it rose, the nuclear power market collapsed. The drop in consumption of electricity as a result of the energy crisis was an ironic turn of events, reducing the need for new plants of any kind. The setback to the nuclear power industry was further aggravated by the break-up of The AEC in 1974 and the division of its promotion and regulatory authorities. The nuclear power industry was also reeling in the wake of an accident at the Browns Ferry nuclear plant near Athens, Alabama, in 1975 and public protests at sites such as Seabrook, New Hampshire (1976). The economic realities of the mid-1970s may have been more significant in undermining nuclear power than the protests, however. True, the accident at Three Mile Island in 1979 (like the disaster at Chernobyl in 1986) severely questioned the viability of commercial nuclear power (Walker, 2004; Cantelon and Williams, 1982; Martin, 1980; Ford, 1982a). In many respects, however, the circumstances surrounding the accident at Three Mile Island was not the departure point for a loss of faith in nuclear power in the United States, but the climax.

Projections about the current and future importance of nuclear power in the energy mix – whether positive or negative – do not detract from several conclusions: (1) nuclear power did not set off an energy transition in the United States; (2) in many respects the development of commercial nuclear power had little to do with energy demand or a concern about declining fossil fuel supplies; and (3) the debate over nuclear power engaged people on a broad societal controversy – not a narrow economic argument. Proponents envisioned a source of energy that would sustain the kind of economic and material development that they perceived as the linchpin of the so-called 'American way'. Opponents saw a risky technology that still retained the characteristics of the older energy sources – concentrated control in the hands of the few – and one that threatened potentially disastrous environmental repercussions. Thus nuclear power in the United States was developed in a manner strongly influenced by the nature of energy transitions in the past, but did not, alone, possess the characteristics of an energy source able to topple dependence on fossil fuels (Ford, 1982b; Mazuzan and Trask, 1979; Del Sesto; Rolph, 1979, Myers, 1977, Aviel, 1982; Burn, 1978).

On a mundane level, the search for alternative energy futures led to a search for a new or little-utilized energy source to replace or complement those that were becoming scarce (or were perceived as scarce). Several untapped sources of fossil fuels were available, including the conversion of coal into gas or liquid, the liquefaction of shale to retrieve oil, the extraction of oil from tar sands, and efforts at tertiary recovery from oil deposits. Various forms of solar energy attracted considerable attention, as did geothermal energy, controlled thermonuclear fusion, the conversion of hydrogen into liquid, and other alternatives. However, the debate over alternatives went beyond tangible issues of replacement of old fuels with new ones. As a Resource for the Future study stated, 'Energy has become the testing ground for conflict over broader social choices'. Several long-held values and traditions came under scrutiny. Issues no less fundamental than individual rights and freedoms, economic equity, the preservation of the environment, the role of government, and world peace were introduced into the debate over America's energy future. The convergence of the environmental movement and the energy crisis made economic growth a focus for debate.

This 1970s discussion about values, traditions and development of course already had deep precedents. In the 1960s population growth had become a highly publicized topic. Kenneth Boulding criticized unrestrained growth as reckless, and pioneered the concept of 'spaceship earth'. A 'spaceship economy' – an economy of limited resources – took into account the finite nature of the world's resources, suggesting a move away from the idea of unlimited growth. This was an early form of 'ecological economics' – or even sustainability – which placed emphasis on social and political equity as well as economic growth (Henderson, 1973; Roelofs, Crowley and Hardasty, 1974; Solberg, 1976; Petulla; Olson and Lansberg, 1973; Ophulus, 1977).

By the 1970s, with the onset of the energy crisis, the debate over growth moved beyond environmental issues to include energy issues. Of particular concern were alternative energy futures for the United States. Proponents of continued energy growth noted the past benefits – comfort, material well-being, high employment, and more leisure time. They maintained a faith in the market mechanism to adjust to scarce resources, to produce technical fixes in exploiting available energy sources, and to create or discover new sources. Opponents of unlimited growth questioned economic practices leading to more material goods, rather than better services or an improved quality of life.

Some critics stressed the social and political consequences of unlimited growth. Complex, centralized systems that produced and sustained growth might threaten personal liberty, even the democratic process. Several major adjustments might be employed to redress these potential and real inequities: steering growth away from resource-intensive industries, redistributing income, giving higher priority to quality-of-life interests. In essence the effort to limit growth would require a shift in values.

Amory B. Lovins – a young American physicist living in Britain, representative of the Friends of the Earth, and ardent opponent of nuclear power – focused the anti-growth debate on the energy issue with his article, 'Energy Strategy: The Road Not Taken?' published in *Foreign Affairs* in 1976. In his various works Lovins presented a critique of contemporary energy systems and offered a radical alternative. Lovins stated that the policy of sustaining growth in energy consumption and limiting imports was no answer at all. Instead, he recommended an 'end-use orientation', that is, to determine 'how much of what kind of energy is needed to do the task for which the energy is desired, and then supplying exactly that kind' (Lovins, 1977). In Lovins' eyes the energy problem was not so much tied to the source as to the society that used it. A significant social change, he reasoned, was necessary to get off the 'hard energy path' and onto the 'soft energy path'. Hard energy paths were 'high-energy, nuclear, centralized, electric'; soft paths were 'lower-energy, fission-free, decentralized, less electrified' (Lovins and Price, 1975; Nash, 1979).

Lovins's dichotomy between hard and soft energy paths attracted considerable attention and controversy. Some energy experts refined the construct or borrowed from it in touting their own energy strategies for the future. Numerous other reports tried to promote alternative perspectives, including 'consensus' positions, 'balanced' energy programs, and so forth (Ridgeway and Conner, 1975; Perelman et al., 1981; Morris, 1982). Advocates of the hard path, or critics of the soft path (these were not necessarily the same), questioned Lovins's defense of 'appropriate' technology and his decentralist position. It was common for critics to characterize the decentralist approach as 'romantic' or simply as promotion for 'post-industrial pastoral society' (Florman, 1981; Petroski, 1982).

Extremes, whether posed by Lovins or his opponents, stimulated debate over alternate energy futures, but did not undercut more modest appraisals, especially by those who tried to emphasize the differences among short-term, intermediate, and long-term energy requirements and goals. In the modern era, energy issues were also worldwide in scope and impact. Some critics could not reconcile a decentralization plan with that fact. Several studies favoring a new direction did not begin with the premise that Lovins's view offered the only hope for the nation. Instead, they argued that United States energy prospects were not likely to change dramatically overnight, and thus goals for the immediate future were most pressing.

The nation faced difficult choices by the 1980s – or so it seemed. The tone of impending doom that engulfed the nation in the 1970s seemed to subside almost as quickly as it rose. In the early 1980s, some commentators were trumpeting the declining consumption of oil and other fuels and the loosening grip of OPEC as signs that the energy crisis was fading. The report of the President's Commission for a National Agenda for the Eighties (1980), appointed by President Jimmy Carter, asserted that the nation's energy predicament was not yet resolved, but, it added, 'Past reliance on the rhetoric of crisis has probably harmed the nation's ability to cope with its energy problems because Americans soon discovered that predictions of imminent catastrophe did not materialize'.

In the aftermath of the energy crisis and in the midst of a cautious optimism about the nation's energy present, conservative Ronald Reagan became the thirty-ninth President of the United States. In 1981, the new president offered the country 'the elixir of "supply-side economics" mixed with a strong draught of military spending' to cure its economic woes (Leuchtenburg, 1983). Reagan's approach to government was built on faith in the productive capacity of the United States – not a fear of future energy shortages.

Within the next few years, oil prices stabilized and even declined as a result of an international oil glut. Experts warned about the transitory nature of the consumers' boom – and the possible financial repercussions – but it was difficult not to breathe a little easier. When the American economy began to climb out of the recession, it seemed that the decade of the 1970s was eons away (Newsweek, 18 May 1981; 14 September 1981). The National Energy Policy Plan, announced on 4 October 1983, formally abandoned the goal of US energy independence set in the 1970s. Energy Secretary Donald Hodel asserted: 'This plan does not contemplate total self-sufficiency. This contemplates working toward what I would call energy non-dependence, in which we continue to import where that makes economic sense, but not to the extent that an interruption undercuts our economy or our military capability' (Houston Post, 5 October 1983).

The return of optimism in the early 1980s may have been temporary, but it suggested that historic forces have had a stronger impact on the nation's energy present than the fading memories of the energy crisis. The United States did not move quickly toward a post-petroleum economy after 1979, and aside from some conservation of gasoline, electricity, and home heating fuels, Americans did not lose faith in America the Abundant.

During the energy crisis, policy makers were obsessed with the role of the United States as a consumer of energy. In the early 1980s, the Reagan administration attempted to swing the pendulum back by focusing on energy production, utilizing the economic hard times to justify more intense energy exploration and to deregulate private energy supplies. The primary force behind the administration's faith in private development of energy was a reaffirmation of the historic role of government as promoter of economic growth. What it rejected was the additional role of regulator and intervener. The Reagan administration's market orientation did not lead to unrestrained competition in the energy field. There was no attempt to tamper with the multinationals and large oil independents or to frustrate oil companies from diversifying into other energy sources – namely coal and nuclear power. The primary goal was to remove governmental impediments to corporate action in the hope of stimulating economic growth. Immediate decontrol of natural gas, for example, was deferred. And the administration jealously guarded the central role of the federal government in the promotion of nuclear power.

In some ways the Reagan administration tried to make good on its pledge to reduce regulatory impediments to energy production and to streamline the federal energy bureaucracy, but gave little attention to long-range energy needs.

Although energy and environment were inextricably linked as issues in the 1970s, the setting for debate changed markedly in the early 1980s. Environmental groups were well entrenched within the economic and political institutions of the United States, but their power and influence was still largely dependent on the tone and actions of government. In the 1960s and 1970s, the federal government may not have always led the way in environmental issues, but it had paid lip service to many environmentalist goals. The notion of 'environmental costs' came to be included in almost every discussion of energy policy. The Reagan administration, however, demonstrated a cavalier attitude toward the environment. In some cases, environmental risks from energy exploration and development were dismissed or ignored. Conquering recession and restoring American world prestige, the President believed, were more important than the wants of over-sensitive preservationists. The appointment of James G. Watt as Secretary of the Interior and Anne McGill Burford (formerly Gorsuch) to head the EPA were clear reminders of where the administration's priorities stood. Watt, in particular, symbolized the commitment to economic growth at all costs (*Newsweek*, 5 January 1981; *Time*, 14 September 1981; 21 February 1983; 28 February 1983; *New York Times*, 27 March 1983; Tirman, 1982).

Despite the return of a short-term view of energy policy under the Reagan administration, signs of a third energy transition did not totally disappear. Energy experts were quick to point out that the nation had experienced a 'fuel' scarcity rather than a full-fledged 'energy crisis' in the 1970s. In an absolute sense, there was plenty of energy to be exploited, but a fuel scarcity was a very serious matter. At least the energy crisis helped to dispel the notion that American energy sources were inexhaustible. It also was becoming apparent that energy-use habits over the years had strongly influenced American culture. Not only had the United States become a petroleum-based society in the twentieth century, but it had become the most energy-intensive society on earth. It was committed to a one-dimensional transportation system dominated by the automobile and utterly dependent on electrical power. Industry was increasingly mechanized and much less labor intensive. Thus, the emerging culture was fragile in key ways despite its apparent strengths. Yet despite the danger signs exhibited in the 1970s and 1980s, and despite a flicker of debate over energy, consumption, corporate profits, and potential scarcities, signs of a major energy transition did not materialize in the twentieth century.

A Hydrogen Society

In the late-twentieth century and into the twenty-first century, attention to past energy transitions was abandoned for concern about potential future ones. Creating a sustainable – or renewable – energy future as a goal gained credibility among those who were dissatisfied with the old notion of America the Abundant. Some believed that the world energy economy was poised for a sweeping shift away from imported oil and environmentally damaging coal in the next several decades. Studies showed that direct combustion of fuels for transportation and heating accounted for two-thirds of the total greenhouse gas emissions. The transition to a 'hydrogen economy' or a 'hydrogen society' – employing hydrogen fuel cells that emit only water and other environmentally friendly technologies – seemed like a promising option. Some commentators suggested that industrialized countries might get started on the hydrogen path by 'piggybacking' on existing energy infrastructures. However, the prospect of a new hydrogen economy are highly uncertain, even though recent projections see a decline in oil resources within a decade or so (Hoffman, 2002; Ogden, Garman, and Huberts, 2003; Rhodes, 2002; Hesse, 2003; 'Renewable energy', *Belfast Telegraph Newspapers Ltd*, May 30, 2003; *Chemical Week Associates* 18, January 20, 2003; *US Newswire*, February 5, 2004; Paulsn, 2003).

In light of prospects, or hopes, for an energy transition in the next several decades and the experience we have had in attempting to understand past transitions, some important questions can be raised:

1.   Is discussion of a transition to a sustainable fuel taking into account changes that will be needed not only in energy sources and energy infrastructures, but also in changing consumption behavior and potentially broad cultural and social transformations?
2.   Which sectors of the world economy will need to retain their dependence on fossil fuels and for how long? In developed countries? In developing countries?
3.   Does a focus on a 'hydrogen society' really equal sustainability? By what measures?
4.   Aside from changes in energy use, what attention needs to be paid to energy conservation?
5.   Are projections, built only on the presumption of maintaining the existing lifestyles, taking into account the cost of such maintenance and the great variety of lifestyles worldwide?

Conclusion

Such difficult questions – and there are many more – suggest that a deeper examination of past energy transitions might be useful in looking toward the future if for no other reason than for exposing the complexity of such changes. If history teaches us anything on this subject it is that transitions are not simply exercises in swapping fuels and changing technologies, but disruptive events with the potential to remake societies in fundamental ways.

— disappointing concl — not really able to address relevancy of past for the present.

References

Allen, W. (1977), *Nuclear Reactors for Generating Electricity: US Development from 1946 to 1963*, v-x, 76-80, Rand, Santa Monica, CA.

Andrews, R.N.L. (1980), 'Class Politics or Democratic Reform: Environmentalism and American Political Institutions', *Natural Resources Journal* 20, April, 221-41.

Aviel, S.D. (1982), *The Politics of Nuclear Energy*, Washington, DC, 30-41.

Balogh, B. (1991), *Chain Reaction: Expert Debate and Public Participation in American Commercial Nuclear Power, 1945-1975*, Cambridge University Press, Cambridge.

Burlingame, R. (1938), *March of the Iron Men: A Social History of Union through Invention*, Scribners, London.

Burn, D. (1978), *Nuclear Power and the Energy Crisis*, Macmillan, London.

Cantelon, P.L and R.C. Williams. (1982), *Crisis Contained: The Department of Energy at Three Mile Island*, Southern Illinois University Press, Carbondale, IL.

Cochran, T.C. (1981), *Frontiers of Change: Early Industrialization in America*, Oxford University Press, Oxford.

Cole, A.H. (1970), 'The Mystery of Fuel Wood Marketing in the United States', *Business History Review* 44, Autumn, 355-356.

Dawson, F.G. (1976), *Nuclear Power: Development and Management of a Technology*, University of Washington, Seattle.

Del Sesto, S.L. (1979), *Science, Politics, and Controversy: Civilian Nuclear Power in the United States, 1946-1974*, Westview Press, Boulder, CO.

Energy Policy Project, Ford Foundation (1974), *A Time to Choose: America's Energy Future*, Cambridge, MA.

Florman, S.C. (1981), *Blaming Technology*, St. Martin's Press, New York.

Ford, D. (1982a), *Three Mile Island: Thirty Minutes to Meltdown*, Penguin, Harmondsworth.

Ford, D. (1982b), *The Cult of the Atom*, Simon and Schuster, New York.

Gillette, R. (1972), 'Nuclear Safety: The Roots of Dissent', *Science* 177, 1 September, 771-76.

Gillette, R. (1972), 'Nuclear Safety: The Years of Delay', *Science* 177, 8 September, 867-71.

Goodwin, C.D. (ed.) (1981), *Energy Policy in Perspective: Today's Problems, Yesterday's Solutions*, Brookings Institution, Washington DC.

Greenberg, D. (1980), 'Energy *Flow* in a Changing Economy, 1815-1880', in Frese, J.R. and Judd, J. (eds), *An Emerging Independent American Economy, 1815-1875*, Sleepy Hollow Press, Tarrytown, NY.

Henderson, H. (1973), 'Ecologists versus Economists', *Harvard Business Review* 51, July, 28-36, 152-57.

Hesse, S. (2003), 'Alternative Power is Set to Blow Away the Old', *Japan Times Ltd*, 14 August.

Hindle, B. (1975), 'Introduction: the Span of the Wooden Age', in Hindle, B. (ed.), *America's Wooden Age: Aspects of Its Early Technology*, Sleepy Hollow Press, Tarrytown, NY.

Hoffman, P. (2002), *Tomorrow's Energy: Hydrogen, Fuel Cells, and the Prospects for a Cleaner Planet*, MIT Press, Cambridge, MA.

Hunter, L.C. (1949), *Steamboats on the Western Rivers*, Harvard University Press, Cambridge, MA.

Keating, W.T. (1979), *Politics, Technology, and the Environment: Technology Assessment and Nuclear Energy*, Arno Press, New York.

Kranzberg, M., Hall, T.A. and J.L. Scheiber (eds) (1980), *Energy and the Way We Live*, Boyd and Fraser, San Francisco.

Landsberg, H.H. and S.H. Schurr (1968), *Energy in the United States: Sources, Uses and Policy Issues*, Random House, New York.

Leuchtenburg, W.E. (1983), *A Troubled Feast: American Society Since 1945*, Rev. ed. Little, Brown, Boston.

Lillard, R.G. (1947), *The Great Forest*, A.A. Knopf, New York.

Lovins, A.B. (1977), *Soft Energy Paths: Toward a Durable Peace*, Penguin, Harmondsworth.

Lovins, A.B. and J.H. Price. (1975), *Non-Nuclear Futures*, Friends of the Earth International, San Francisco.

Martin, A. (1976), 'James J. Hill and the First Energy Revolution: A Study in Entrepreneurship, 1865-1878', *Business History Review* 50, Summer, 179-197.

Martin, D. (1980), *Three Mile Island: Prologue or Epilogue?*, Cambridge, MA.

Mazuzan, G.T and R.R. Trask (1979), *An Outline History of Nuclear Regulation and Licensing, 1946-1979*, Historical Office, Office of the Secretary, Nuclear Regulatory Commission, Washington DC.

Mazuzan, G.T. (1981), 'Conflict of Interest: Promoting and Regulating the Infant Nuclear Power Industry, 1954-1956', *Historian* 44, November, 1-14.

Melosi, M.V. (1985), *Coping with Abundance: Energy and Environment in Industrial America*, Knopf, New York.

Melosi, M.V. (1986), 'The Third Energy Transition: Origins and Environmental Implications', in Bremner, R.H, Reichard, G.W. and Hopkins, R.J. (eds), *American Choices: Social Dilemmas and Public Policy since 1960*, Columbus, Ohio, pp. 187-218.

Morris, D. (1982), *Self-Reliant Cities*, Sierra Club Books, San Francisco.

Myers, D. III (1977), *The Nuclear Power Debate*, Praeger, New York.

Nash, H. (ed.) (1979), *The Energy Controversy: Soft Path Questions and Answers*, Friends of the Earth, San Francisco.

Nettels, C.P. (1962), *The Emergence of a National Economy, 1775-1815*, Holt, Reinhart and Winston, New York.

Nye, D. (1998), *Consuming Power: A Social History of American Energies*, MIT Press, Cambridge, MA.

Ogden, J.M., Garman, D.K. and D.P.H. Huberts (2003), 'Transition to a Hydrogen Society', Capitol Hill Hearing Testimony, Washington, DC: Federal Document Clearing House, Inc., 5 March.

Olson, M. and H.H. Lansberg (eds) (1973), *The No-Growth Society*, Norton, New York.

Ophuls, W. (1977), *Ecology and the Politics of Scarcity*, W.H. Freeman, San Francisco.

Paulson, T. (2003), 'NW is Poised for Hydrogen Fuel Role', *Seattle Post-Intelligencer*, B1.

Perelman, L.J. (1981), 'Speculations on the Transition to Sustainable Energy', in Perelman, L.J, Giebelhaus, A.W. and Yokell, M.D. (eds), *Energy Transitions: Long-Term Perspectives*, Westview Press, Boulder, CO, 185-213.

Perelman, L.J., Giebelhaus, A.W. and Yokell, M.D (eds) (1981), *Energy Transitions: Long-Term Perspectives*, Boulder, CO.

Perry, R. (1977), *Development and Commercialization of the Light Water Reactor, 1946-1976*, Rand Corporation, Santa Monica, CA.

Petroski, H. (1982), 'Soft Energy Technology is Hard', *Technology Review* 85, April, 39.

Petulla, J.M. (1977), *American Environmental History: The Exploitation and Conservation of Natural Resources*, San Francisco, 102-103, 122-124.

Petulla, J.M. (1980), *American Environmentalism: Values, Tactics, Priorities*, Texas A and
   M University Press, College Station.
Pratt, J.A., 'The Ascent of Oil: the Transition from Coal to Oil in Early Twentieth-Century
   America', in Perelman et al. (eds), *Energy Transitions*, 9-29.
Rhodes, R. (2002), 'Energy transitions: a history lesson', *Center for Energy Research
   Newsletter*, 2 June.
Ridgeway, J. and B. Conner (1975), *New Energy*, Beacon Press, Boston.
Roelofs, R.T., J.N. Crowley and D.C. Hardasty (eds) (1974), *Environment and Society: A
   Book of Readings on Environmental Policy, Attitudes and Values*, Prentice-Hall,
   Englewood Cliffs, NJ.
Rolph, E.S. (1979), *Nuclear Power and Public Safety,* Lexington Books, Lexington, MA.
Schurr, S.H and B.C. Netschert (1960), *Energy in the American Economy, 1850-1975*, Johns
   Hopkins University Press: Baltimore, 45-124.
Schurr, S.H. et al. (eds) (1979), *Energy in America's Future: The Choices Before Us*, Johns
   Hopkins University Press: Baltimore.
Solberg, C. (1976), *Oil Power*, New York, 222-25.
Stobaugh, R. and Yergin, D. (eds) (1979), *Energy Future: Report of the Energy Project at
   the Harvard Business School*, New York.
Tarr, J.A. and B.C. Lamperes (1981), 'Changing Fuel Use Behavior and Energy Transitions:
   The Pittsburgh Smoke Control Movement, 1940-1950', *Journal of Social History* 14
   (Summer), 561-588.
Taylor, G.R. (1951), *The Transportation Revolution, 1815-1860*, New York, 56-73.
Tillman, D.A. (1978), *Wood as an Energy Source*, New York, 11.
Tirman, J. (1982), 'Investing in the Energy Transition: From Oil to What?', *Technology
   Review* 85 (April), 65-72.
Walker, J.S. (2004), *Three Mile Island: A Nuclear Crisis in Historical Perspective*,
   University of California: Berkeley.

Internet Sources

'The Hydrogen Society', www.teknologiradet.no/html.

Newspaper Articles

'Renewable energy' (30 May 2003), *Belfast Telegraph Newspapers Ltd*.
'DOE gazes into the post-petroleum future' (20 January 2003), *Chemical Week Associates 18*.
Houston Post (5 October 1983), D-1.
'Hydrogen can fundamentally transform the US system' (5 February 2004), *National
   Academies' Report, US Newswire*.
Newsweek (18 May 1981), 32-33; (14 September 1981), 74-75; (5 January 1981), 17.
New York Times (27 March 1983), E5. 55.
'Transitioning to a renewable energy future' (17 November 2003), *M2 Communications Ltd*.
Time (14 September 1981), 18; (21 February 1983), 14-16; (28 February 1983), 17.

Chapter 2

Hydraulic Energy, Society and Economic Growth

Salvatore Ciriacono *Hstoy, Padua*

Numerous studies have underlined how, throughout history, the economic development of cultures is bound up with the control of energy resources. Such a connection is now accepted as beyond reasonable doubt. However, when one *rather broad* speaks of energy and its uses one is also referring to technology, scientific knowledge and man's capacity to harness what is required for further development of economic activities. Hence, energy, technology and science seem to form a single whole – and that is the perspective I will adopt in looking at the use of water-generated energy during the course of history, with the aim of outlining the existence of phases of rapid expansion and stasis, interspersed with long periods of transition. During these latter periods we observe the search for alternative energy *long term dynamics* sources accompanying (or causing) an economic stagnation and social malaise, which are only overcome once the possibility of new energy supplies begins to *expansion/stasis/transition* take real form.

Obviously, this approach does not entirely convince those who see economic decline as being explained by other causes – for example, agricultural productivity, the 'law of diminishing returns', population size, the inability to properly exploit available resources, the decline of shared values or even social conflict. Naturally, none of these causes can be ignored and one has to establish the links between the *multi-variate* different variables that come into play in a given situation. For example, whilst it is undoubtedly true that throughout history energy resources have ultimately proved to be limited – and insufficient to sustain economic development – it is also true *not at all evident to me* that the search for alternative energy sources and the resulting development of know-how in a particular national, social or professional context (that is, the nurturing of specific human capacities) has been the *conditio sine qua non* for overcoming the socio-economic regression that has been produced by such energy difficulties.

The points I would like to examine with regard to the exploitation of water power concern: 1) the relation between energy and economic growth; 2) the development of scientific knowledge and the exploitation of energy resources in any given period; 3) the competition for energy resources which develops over time between various social groups (at the intra-social, national or international level).

Energy and Economic Growth

Though perhaps excessively categorical, the best summary of the role energy plays in economic development or decline is the following statement by J.W. de Zeeuw: 'in a given period the greatest prosperity (with opportunities for the advancement of technological and social attainments) arises in the area, where – depending on geophysical circumstances governing the winning and transport of energy – the acquisition and application effort (in the light of already available skill) per serviceable energy-unit is the smallest. The economic decline of an area is inevitable as soon as these circumstances become relatively less favourable in that area' (de Zeeuw, 1978, p. 26). Whilst such a conclusion would seem to argue that scientific know-how – the so-called 'knowledge society' – plays a secondary role, it also makes clear that geographical location and the concomitant availability of energy (for example, water) are far from negligible factors. This question is best illustrated by a comparison of what happened in the Middle East and the West in the period that runs from the Middle Ages to the Early Modern period. Though developed in the Mediterranean area just before the Christian era, the watermill would play its fullest role in economic development in Western Europe as a whole; due to the inevitable uncertainty of water supplies in the countries of the Middle East, it was there a far less fruitful generator of energy and economic growth. As C. Issawi has pointed out, the Doomsday Book of 1086 reveals that 'the 3,000 settlements, most of them very small, contained 5,624 mills, some of which were used for other purposes than grinding corn (…) In the next three centuries, everywhere in Europe watermills were put to a variety of uses, in the textile, metallurgical, wood, and other industries, and they stimulated a number of inventions or borrowings, such as the cam and the crank. In the 16th century, at the height of Ottoman prosperity, the Middle East had distinctly fewer mills per capita than 11th century England' (Issawi, 1991, pp. 284-5). Clearly this situation played its part in the relative decline of the Middle East when compared to the West; the latter would, right up to the Industrial Revolution, go on increasing the number of both its watermills and windmills.

The studies published by Eliyahu Ashtor show how at the end of the Middle Ages a long-standing 'divergence' between Western Europe and the Islamic world had become established (Ashtor, 1976), even if in previous centuries the latter had long been more advanced than Europe in various areas of science, technology and craft skills. And in this change-around, energy resources were a key factor. Excluding manpower and animal labour, wind, wood and water would, right up to the advent of the steam engine, be the bases of what has been described as an 'eotechnic economy' (Lewis Mumford and Patrick Geddes) or, more recently, as an 'advanced organic economy' (E.A. Wrigley, 1988); and there is no doubt that, for geographical and environmental reasons, the Middle East is at a disadvantage where all three are concerned. In saying this, I do not want to underestimate the detrimental effects of a certain traditionalism in Arabic science, which has been stressed by Joel Mokyr and other scholars of the 'Technological School'; with their focus on such notions as 'exposure effect' and 'path dependency' (Mokyr, 1992), the studies produced by these historians cast light how scientific knowledge (which today goes to make up the 'knowledge society' and is considered part of human capital) had a decisive effect upon the shortfall between Arabic technological culture and that in the West. Nevertheless, if pursued too narrowly, this interpretation would underestimate the importance of the Middle East's limited energy resources – and not only in water, wood, wind: look, for example, at England, which was already exploiting its coal deposits by the sixteenth century and thus enjoying a sort of Pre-Industrial Revolution. Obviously, there is also the danger of going too far in the opposite direction, trying to explain and measure the crisis within the *ancien régime* solely in terms of energy factors. Whilst there is no arguing with Marc Bloch's now-classic argument that the increasing numbers of windmills played a fundamental role in the economic growth of medieval Europe from the beginning of the year 1000 onwards, the crisis the continent went through in the fourteenth century cannot be explained away solely in terms of energy resources; other key factors that have to be taken into account in any convincing discussion include epidemics (the plague), stagnant agricultural yield ratios and demographic pressure. In effect, this crisis in the economy of fourteenth-century Europe has quite aptly been defined as an 'environmental' one which was largely due to excessive exploitation and consequent washing-away of terrain (Crescenzo de'Crescenzi), unsustainable population growth-rates and a demand for natural resources that outstripped supply.

Theme of the chapter is role of water power

The Role of Water Power

The subsequent period of growth in the 15th-17th centuries would see traditional energy sources being exploited in various economic sectors, with the availability of such resources playing a key part in deciding the location of settlements (industrial or otherwise). Certainly there is no need to reiterate how the presence of rivers and watercourses was a pre-condition for the location of numerous urban settlements in the Middle Ages and in more ancient periods of history. The flow of water both within and without medieval cities not only had an effect on the urban fabric (and the creation of city walls enclosed with moats) but it also stimulated the development of manufacturing activities of great economic significance; one need only think of the example of Provins in Northern France. In Italy itself, the Central-Northern regions were home to many such manufacturing cities – the case of Bologna is a perfect example (Guenzi, 1993) – which have quite rightly been described as possessing 'industrial' canals that, even if on a smaller scale, anticipated the role such watercourses would play in the Industrial Revolution. Waterwheels were of key importance in pre-industrial manufactories – as one can see from the ample description given of them in a vast number of Renaissance treatises, which were as precise in technical description and drawing as they were *Boring* beautiful in presentation. As examples of such works, one might mention Vittorio Zonca's *Novo teatro di machine et edificii* (Zonca, 1607), which is one of the most important of the period for its description of the innumerable water powered machines then at work in Italy (not only in that most crucial of all pre-industrial uses, flour-milling, but also in fulling and the production of paper and iron tools). Another similarly representative work is Georg Bauer Agricola's *De re metallica*

(Basel, 1556), which contains numerous illustrations of waterwheels and pumps. Although there is no question that the predominant area of economic activity remained agriculture, these machines played an essential role in a manufacturing sector that, within the more advanced areas of the continent, was becoming no less important. And when one wishes to assess the contribution that water energy made to the total energy produced in pre-industrial economies one has to bear in mind that the former brought with it a qualitative as well as a quantitative change (a fact that does not seem to have been adequately emphasised by the literature on the subject). What is more, in this pre-statistical age, the percentages given for the various energy sources used by man are often rather shaky, being based on partial data and estimates which are extrapolated to the extent that they produce rather perplexing results. For example, the following table gives the percentages proposed for the daily energy per capita drawn from the various available energy resources in the period around 1750.

Table 2.1 Daily calories produced by various energy sources

Energy sources	per person	%
Wood and coal	7,200	50
Water and wind	150	1
Animals	5,000	35
Men	2,000	14
Total	14,350	100

Source: P. Malanima, *Energia e crescita nell'Europa preindustriale*, Rome, pp. 119 and 124

[handwritten annotation: What is the point of this chapter? To argue for a cheery role for water?]

Now it is true that in some areas of Europe wind and water might have made a more limited contribution to the daily calorie supply, but it seems far too extreme to reduce it to as little as 1 percent. The main objection one could make to such figures is that they take into account the role of wind and water in producing flour (which obviously played a role in the daily calorie supply), but totally ignore their role in the manufacturing sectors which made a far from negligible contribution to that supply. For example, were not wind and water power used extensively in the metallurgy that then made its contribution to the construction of mills, machinery and gear mechanisms? And though another type of calculation, covering the horsepower generated by various energy sources in late-eighteenth-century Europe, does give greater importance to water power, putting its contribution at some 12 percent of the total seems to underestimate the real importance of this energy resource. According to Fernand Braudel an individual man was not 'a very powerful engine. His strength measured in horsepower (seventy-five kg raised to a height of one metre in one second) is derisory: between 3-4 percent of one horsepower against 27 to 57 percent for a cart horse. In 1739, Forest de Belidor maintained that seven men were required to do the haulage work of one horse' (Braudel, 1982, p. 353). It is calculated that in the 18th century 14 million horses and 24 million cows – 38 million animals – produced about 10,000,000 horsepower, 'each animal representing a quarter horsepower, an important contribution to the continent's power supply' (ibid.). Given that the population of eighteenth-century Europe is estimated at c.150 million, this relation of four people per work/transport animal seems about right. Every day, the food consumption of each work animal was around 20,000 calories – that is, more or less, the daily food intake of seven men (Malanima, 1996).

Table 2.2 Energy production in the preindustrial world

	Millions of HP	%
Animals	10	41
Men	0.9	3
Wood	10	41
Waterwheels	1.5-3	12
Sails	0.2-3	1

Source: Braudel, 1981; see also Sieferle, 1987

Furthermore, one also has to bear in mind that not all the energy used is transformed into useful energy – that is, the yield in any form of energy conversion is always less than 100 percent. Expressed as a formula, one might give energy (e) in these terms:

$$e = W/Q$$

where W = work and Q = production of energy. When heat energy is being transformed into motion, the yield is particularly low, with sizeable loss of energy; however, the kinetic energy of water can be transformed into mechanical energy (in watermills etc.) with relatively low loss, given the limited friction at the waterwheel. The following table for such conversions has been drawn up:

Table 2.3 Energy production and yield

Source of energy	Yield ratios, %
Wood	25
Water and wind	35
Animals	10
Men	20

Source: Malanima, 1996, p. 120

And whilst it is true that the power of a waterwheel (the power (P) generated by a waterwheel was directly proportional to the kinetic energy of the water and the distance it fell $P=HW$, where $W=$Flow and $H=$Slope) is lower than that generated by a windmill. If one looks at Europe as a whole (setting aside the, admittedly very innovative, Netherlands) one sees that the former were much more widespread and therefore more important. For Braudel 'the power from these "primary engines" was probably not very great – from two to five horse-power from a water-wheel, sometimes five, at most ten, from the sails of a windmill' (Braudel, 1982, p. 353). De Zeeuw has calculated that in the seventeenth century a windmill could generate 60 Kw, equivalent to the work output of 100 men (De Zeeuw, 1978). Here again one might refer to Braudel, who comments: 'towards the end of the eighteenth century there were, according to records for Galicia (which had come under Austrian rule), 5243 watermills and only 12 windmills in an area of 2000 square leagues and for a population of 2 million (…) If the ratio of watermills to population was the same elsewhere as in Poland, there would have been 60.000 in France and not far off 500.000 to 600.000 in Europe on the eve of the industrial revolution (…) As a rule, every village had its mill. Where there was not enough running water, on the Hungarian plains for instance, mills were operated by horses or even manual labour' (one should not forget here that 'the yield of a hand mill operated by two men was only one fifth that of a watermill …') (Braudel, 1982, pp. 357, 358, 371).

Water and Other Energy Sources

While stressing how the importance of water power in the pre-industrial ages of Europe and elsewhere has been partially undervalued, I certainly do not wish to diminish the role played by other energy sources. If one looks at the Netherlands – which, with some overemphasis, has been described as home to the 'first modern economy' – one sees the importance of both wind and peat. Starting with Simon Stevin, wind was studied as scientifically as water flow, thereafter being harnessed by ever more sophisticated windmills that drove machines which made an essential contribution to the economic and industrial success of the Netherlands. It was these which made it possible for the country to surpass its various rivals, at first Italy and then England: the former continued to pin its increasingly meagre economic fortunes on the exploitation of water power; the latter – even if it had already begun in the sixteenth century to make more massive use of 'sea-coal' than any other European nation – had yet to be in the position it would gain in the seventeenth century, when conditions were ripe for the 'take off' of industrial activity.

Nevertheless, even if key factors in the Dutch success story, these forms of 'organic and renewable' energy (E.A. Wrigley, 1988) were suited to a certain phase in the history of manufacturing industries. Over time they would prove to be not only in short supply (triggering what we know as 'energy crises'), but also incapable of meeting the demand resulting from an ever-growing population and ever more radical transformations of the production process. With regard to wind, the limits most commonly stressed were its very unpredictability, i.e., the fact that energy supply could not be adjusted to meet production requirements. Furthermore, though there is no doubt that the windmill and efficient harnessing of wind power would ultimately become the very symbols of Dutch economic achievement, it should not be forgotten that in the Low Countries this energy source was closely linked with the industrial sector as well with the agricultural: the traditional windmill was adapted to the needs of land reclamation, with the wheels raising water to drain water-logged terrain (over a period from the 15th to the 18th century, these *wipmolen*, drainage mills, would undergo continuous improvements – for example, in the 17th century the traditional wheel was replaced by an Archimedes' Screw, thus making it possible to drain even lower terrain).

As for peat, its role in the growing international and economic status of Holland is not to be underestimated. As has been stressed: 'this Golden Age was based on the easy availability of a huge amount of peat covering the domestic need for energy, just as the British Empire was based on coal and the present dominance of the United States on the cheap supply of oil and natural gas' (De Zeeuw, 1978, pp. 25-6). Yet it has also opportunely been pointed out that though 'this thesis is not incorrect, it reduces the prosperity of the Low Countries to only one aspect', whereas we know that a full interpretation has to take the play of various factors into account (Stol, 2004). What is true is that, though both wind and peat continued to be used as sources of energy in the 17th and 18th century, the Dutch economy still went into a phase of stagnation and decline. The reasons for this deterioration are numerous and cannot be reduced solely to the question of energy supplies – even if there is no denying that these did play a part and that the concomitant energy 'crisis' is revealed by the shift in energy expenditure towards coal and steam. With regard to drainage wheels, the introduction of steam power certainly marked 'a decisive turning-point that made it possible to finally establish control over key areas of drainage and land reclamation not only in the Low Countries but in the whole of Europe' (ibid.). In the Low Countries the large area of the Haarlemmer-meer would not be drained until the nineteenth century, using steam-driven pumps; every attempt up to then had failed. The first modern steam pump was tried experimentally near Mijdrecht in 1794 (E. Schultz, 1992). It has been observed, for example, that by 1852 'in the Fens, the number of windmills formerly at work between Lincoln and Cambridge' had dropped from 700 to 220 (Darby, 1956). With steam-powered drainage it became possible to increase not only the area of land drained but also the cost efficiency of the entire operation (though, of course, sizeable initial investment was required; it is no coincidence that the use of the steam pump meant the setting-up of sizeable consortia that could draw on adequate financial backing).

Slowly but surely, the mystique of the windmill and the waterwheel fell victim to the advance of steam engines and coal (especially coke, which was so closely associated with such engines) (Greenburg, 1990).

Energy in Transition

However, as E.A. Wrigley so rightly points out, the shift from an organic fuel to an inorganic (or mineral) fuel economy was not as sudden as one is led to believe by those historians anxious to focus upon the glories of the Industrial Revolution. This is particularly true in the case of water power. Throughout the 18th and 19th centuries, the applications and productivity of this energy source continued to increase, and it was of fundamental importance in the arrival of modern industrialisation to many regions.

It is essential to bear in mind that if steam did finally win out over water power, this was not a spontaneous, inevitable process; the force with which the logistical and economic reasons for this victory made themselves felt varied from one area of developing industrial activity to another.

In the proto-industrial phase, exploitation of water depended on environmental and geographical factors – that is, it gave an advantage to those hilly and mountainous areas that were endowed with free-falling watercourses (it is no accident that it was these areas that became the birthplace of industrialisation). Then, with the advent of the steam engine, which could be freely moved from one place to another, the entrepreneur was no longer bound by such environmental-geographical considerations. However, the fact that, even at this stage, water was not abandoned as a source of energy cannot be explained by saying it had been exploited since time immemorial and was therefore easy to use. Water continued to prove itself a very flexible energy source, capable of increasing levels of efficiency (levels which compared very favourably with those of the first steam engines). During the second half of the 18th century, for example, the traditional waterwheels were joined by the 'breast wheel', developed by John Smeaton in the period 1750-60; this fed the water towards the central part of the wheel (hence its name), thus generating greater kinetic energy than the traditional models which had exploited the pressure of falling or flowing water upon the paddles of a wheel (known, respectively, as the 'overshot' or the 'undershot' variety). It has been calculated that if the 'overshot' wheel could convert two-thirds of total potential energy, by the middle of the nineteenth century innovations had raised that figure to 75 percent (considered a 'ceiling' for a water powered wheel). Other small-scale inventions that made waterwheels ever more efficient included the use of iron for certain components and John Rennie's 1780 design for a 'sliding shutter', which fed the water right to the central part of the wheel even when the level in the channel was above or below optimal depth. The continuing importance of such energy sources is revealed by the fact that Richard Arkwright used water power in his prototype spinning mill at Cromford in Derbyshire (1771), a facility that most historians consider as marking a fundamental step in the advance of the Industrial Revolution itself.

Such observations help one to understand why, around 1800, more energy was being generated in Great Britain by waterwheels than by steam engines – in fact, the former were actually increasing their output. It is no coincidence that towards the end of the 18th and throughout the first decade of the 19th century the price of water in many areas of Britain – particularly the uplands of Lancashire and Yorkshire – increased considerably, with water power often being used alongside the more recently-introduced steam engines. For example, the Sheffield cutlers were in 1794 still using some 111 waterwheels – it was this power source which had initially stimulated the growth of this industry in the city – but only 5 steam-driven generators. The situation in Birmingham does not appear much different; indeed, one – perhaps rather exaggerated – interpretation has it that any 'industrial revolution' there was made possible by the use of water power (this provided the energy for a whole series of manufactories engaged in the production of knives and other bladed instruments). In various areas of the British Isles, sizeable water plants were created – not only to power the machinery of industry but also to meet the water consumption needs of urban centres (for example, Shaw's Waterworks at Greenock in Scotland, a system of reservoirs and water pipes created from 1824 onwards) (Reynolds, 1983; Temin, 1966; Ciriacono, 2000). In short, as Von Tunzelmann has shown 'the social savings in 1800 resulting from the use of steam engines in place of water power in Great Britain were very low, only about 0.2 percent of the national income, equivalent to about one month's growth of the national product at that time' (Tunzelmann, 1978, quoted by Gordon, 1983, no 2, p. 243).

What is more, it is clear that the time steam took to establish its supremacy as a source of energy varied from economic sector to sector: its triumph was relatively quick in the cotton industry, slower in the wool industry and slower still in the metallurgical sector. In effect, with energy demands increasing and the availability of water power being in inverse proportion to its cost, English entrepreneurs had three alternatives: they could move to areas where there was less competition for water, stay where they were but pay more for their water rights, or decide to switch to steam power.

Obviously, only the larger manufactories could initially make this investment, and thus many smaller concerns remained bound to water power for a long time to come; the data at our disposal makes it clear that when a manufactory decided to switch to steam, it had to prepare itself to face a dramatic increase in costs. And even for the larger manufactories, these costs were only acceptable if the more advanced – but more expensive – form of energy resulted in a clear increase in productivity. It was the growing concentration of such manufactories within modern-style factories that made this changeover possible both in Britain and in the countries that industrialised later, where – on the whole – the process was rather less complex.

When one looks at such 'late-comers' one notes the continuing viability of water power as an energy source. For example, in North America, Lowell in Massachusetts – considered one of the 'historic' centres of the industrial revolution within the United States – developed from 1820 onwards largely thanks to massive consumption of water power. And the picture is much the same when one looks at areas that followed Lowell's example: Lawrence (again in Massachusetts) and Manchester in New Hampshire – all three on the river Merrimack. The basic model of industrial development here relied on the creation of a central canal (necessary for the transport of raw materials and finished goods), linked to a series of secondary canals serving the various manufactories which Boston entrepreneurs began to establish from 1820 onwards. By the 1840s the proprietors of manufactories in Lawrence and Lowell were beginning to use locks and dikes to harness the water resources of the natural lakes fed by the Merrimack.

For a long time, the water systems created at Lowell, Manchester and Lawrence would be the most advanced of their kind in the USA, but similar systems were set up along the channels and rivers of America. By 1880 there were about 50 of them, which produced about as much hydro-generated power as the integrated system in Greenock.

As far as steam power is concerned, the differences in the early stages of industrialisation between America and Britain become clear when one considers that 'before Independence only one manufacturing establishment in America was using a steam engine, whereas in Britain perhaps as many as 130 engines were then at work' (Atack, Bateman, Weiss, 1980, p. 281). However, though it is certainly true that in America the very abundance of water supplies initially worked against the introduction of much more costly steam engines, the situation would be reversed during the course of the second half of the nineteenth century: 'by 1900 almost 156,000 steam engines were at work in American factories, where they outnumbered waterwheels and water turbines by four to one' (Atack, Bateman, Weiss, 1980, pp. 281-2). Still, compared to other sources of energy, water power would throughout the twentieth century continue to play a more important role than it did in Europe. For example, it has been calculated that in 1925, 47 percent of the world's hydro-generated power was in America, as opposed to 43 percent in Europe. And this difference would continue thereafter, given that the USA can draw on much more abundant water resources.

When one looks at the improving performance of hydro-electric power – both in those European countries of later industrialisation (France, Italy, Switzerland, Norway) and in North America itself – one sees that one has to shift the chronological divide between an 'organic' and a 'mineral/inorganic' economy. In effect, the 'mineral-based energy economy' did not become fully established until 1870-1914, when coke power was joined by petrol. However, even then, one cannot speak of water power becoming merely a residual energy source. As one historian has observed, this latter was 'one of the oldest utilities (expanding from a bigger base) (…) and (though) the water industry showed one of the least rapid growth rates, nevertheless the expansion was impressive enough'. For Germany and England alone, it has been calculated that there was 'a 15- to 17-fold increase in the aggregate value of water production between 1855 and 1913' (Fenoaltea, quoted by Hassan, 1985, no 4, p. 531).

Conclusion

What is beyond question is that the new sources of power (petrol, gas, coal, electricity and atomic energy) resulted in increases in the standard of living that would have been impossible with traditional energy sources. In effect, energy consumption per capita in Europe would double during the course of the nineteenth century and triple between 1900 and 1989. As the following table shows, the contribution made by traditional energy sources has, over the last few decades, gone into sharp decline.

Table 2.4 Contribution of traditional energy sources to total energy

	Daily calories per person	% of traditional sources
1750	14.350	92
1800	14.750	87
1830	15.150	80
1900	37.590	25
1950	47.430	15
1970	89.560	5
1989	106.700	5

Source: Malanima, 1996

One should also note the increase in productivity and in individual and collective incomes that new energy sources have made possible over the last couple of centuries. From the table below one can see that per capita energy consumption (in calories) increased seven-fold over the period 1820-1929, whilst the increase in average per capita income was fourteenth-fold (where Q= Energy demand and Y= revenue, and k is a constant, Q= kY).

Table 2.5 Per capita energy consumption, 1820-1989

	Yearly revenue ($)		Energy (calories per day and per person)	
1820	1.034	100	15,000	100
1950	4,092	474	47,430	316
1989	14,413	1,393	106,700	711

Source: Malanima, 1996

In certain areas more than others, 'efficiency in the use of different energy sources doubles' (Malanima, 1996); and the overall picture reveals the complexity of the relation between the availability of such resources and the technological and scientific know-how required to exploit them to the full. Here one sees the full economic importance of scientific, cultural and political factors.

References

Ashtor, E. (1976), *A Social and Economic History of the Near East in the Middle Ages*, Collins, London.

Atack, J., Bateman, F. and Weiss, T. (1980), 'The Regional Diffusion and Adoption of the Steam Engine in American Manufacturing', in *The Journal of Economic History* 40, no 2.

Braudel, F. (1982), *Civilization and Capitalism, 15th-18th Century*, vol. 1: *The Structures of Everyday Life*, Harper and Row, New York.

Ciriacono, S. (2000), *La Rivoluzione industriale. Dalla protoindustrializzazione alla produzione flessibile*, Mondadori, Milan.

Darby, H.C. (1956), *The Draining of the Fens*, 2nd ed., Cambridge University Press, Cambridge.

De Zeeuw, J.W. (1978), 'Peat and the Dutch Golden Age', in *AAG Bijdragen* 21.

Fenoaltea, A. (1982), 'The Growth of the Utilities Industries in Italy, 1861-1913', *Journal of Economic History* 42, 601-627.

Giemzo, A. (1993), *Acqua e industria a Bologna in Antico Regime*, Giappichelli, Turin.

Gordon, R.B. (1983), 'Cost and Use of Water Power during Industrialization in New England and Great Britain: a Geological Interpretation', in *The Economic History Review* 36, no 2.

Greenberg, D. (1990), 'Energy, Power, and Perception of Social Change in the Early Nineteenth Century', in *The American Historical Review* 95, no 3.

Hassan, J.A. (1985), 'The Growth and Impact of the British Water Industry in the Nineteenth Century', in *The Economic History Review* 38, no 4.

Issawi, C. (1991), 'Technology, Energy, and Civilization: Some Historical Observations', in *International Journal of Middle East Studies* 23, no 3.

Malanima (1996), *Energia e crescita nell'Europa preindustriale*, La Nuova Italia scientifica, La Nuova Italia, Rome.

Mokyr, J. (1992), *The Lever of Riches: Technological Creativity and Economic Progress*, Oxford University Press, Oxford.

Reynolds, T.S. (1983), *Stronger than a Hundred Men: A History of the Vertical Water Wheel*, Baltimore: Johns Hopkins.

Schultz, E. (1992), *Waterbeheersing van de Nederlandse droogmakerijen*, Lelystad.

Sieferle, R.P. (1987), 'Energie', in F.J. Bruggemeier and T. Rommelspacher (eds), *Besiegte Natur: Geschichte der Umwelt im 19. und 20. Jh*, Munich: C.H. Beck.

Stol, T. (2004), 'Fuel Extraction and Environmental Destruction: Peat and Peat-Companies in the Low Countries (16th-18th centuries)', in *Eau et developpement dans l'Europe moderne*, S. Ciriacono (ed.), Paris.

Temin, P. (1966), 'Steam and Water Power in the Early Nineteenth Century', in *Journal of Economic History* 26, no 2.

Von Tunzelmann, G.N. (1978), *Steam Power and British Industrialisation to 1860*, Clarendon Press, Oxford.

Wrigley, E.A. (1988), *Continuity, Chance and Change: the Character of the Industrial Revolution in England*, Cambridge University Press, Cambridge.

Zonca, V. (1607), *Novo teatro di machine et edificii*, Padua.

Chapter 3

Work and Environment in Mediterranean Europe

Giuliana Biagioli

[handwritten annotations: "Pisa, history" and "highlighting land degradat in Mediterranean"]

The purpose of this chapter is to indicate the sources of energy available to the Mediterranean agricultural sector, and the use made of them, during the period from the end of the Middle Ages down through the twentieth century, as traditional agriculture disappeared. The intent is also to demonstrate that the risk of environmental degradation connected with human activity is not a peculiarity of industrial society, but that it had already manifested itself at the beginning of the period taken into account. Farmers, agronomists and governments strove to minimize this degradation which, had it been irreversible, would have caused the disappearance of entire communities. In fact, the risk of irreversible degradation has increased since the Industrial Revolution and especially during the course of the twentieth century. Only recently have we begun to understand the dangers that this creates for the planet. The dilapidation of the soil has been occurring for thousands of years but now it has increased dramatically. Approximately 35 percent of the earth's soil is degraded permanently.

[handwritten margin notes: "focus + aim", "present danger", "primary human labour"]

During the long period taken into consideration and until the beginning of the twentieth century, the sources of energy used in Mediterranean agriculture were primarily human labour, accompanied by animal labour, supplemented by hydro energy and wind power. In conjunction with these types, energy supplied by wood was also available.

The forest was integrated in the agricultural system and its history is connected with that of farm plots. When agricultural spaces advance, due to demographic and/or economic growth, the forest recedes.

We intend to demonstrate that the use of the energy resources depended:

- On natural factors (the consistent presence of mountains and hills that for a long time hindered agricultural mechanization);
- On environmental conditions (the unhealthy situation of the plains during the entire modern age);
- On economic factors (such as the consistent presence of cities with their necessities of food supply, raw materials, etc);
- On demographic factors: such as the great demographic density of the area in comparison to the rest of Europe, and the consequent low price of labour as a production factor.

[handwritten note: "rather obvious + uninteresting"]

Between the late nineteenth and twentieth centuries, agricultural machinery became widely diffused; however the use of mechanization was confined to a few plains. The hillsides and the mountains remained subject to the labour of humans, aided by the few animals that it was possible to raise, and the small amount of forest still remaining on marginal lands. The diffusion of industrialization and the abandonment of the countryside by the farmers marked the end of an era, of a system of productive and environmental organization, and also the end of an energy system, that was capable of saving land and capital, of recycling every kind of production waste, and that functioned for at least a thousand years.

[handwritten margin note: IMPT / understandit / as / the aim / of / sustainable / development]

[handwritten note: — challenge to today.]

The Original Characteristics

The Environmental Characteristics of Human Activity

According to Fernand Braudel, the Mediterranean is made of earth, sea and air (Braudel, 1972). The air of the Mediterranean is affected by what happens elsewhere: by the Atlantic in the west and by the Sahara in the south. Itself resulting from the sum of microclimates, the Mediterranean unifies them into one climate. The main drawback of the Mediterranean climate derives from the unequal distribution of yearly rains. Rainfall is concentrated in the autumn and spring. The summer drought dries up springs and hinders vegetation. Plants and cultivation systems have taken this factor into account for centuries if not millennia. Water would not be lacking due to the mountains nearby. However, due to the steep nature of the soil, rivers are really like torrents, even a summer storm can cause them to swell enough to damage the farmlands. In a normal year two critical periods alternate: that of sudden floods, which can bring about the destruction of bridges, temporary passes and the flooding of fields, and the period of drought, which drains the beds of torrents and small rivers. Floods mean almost always loss of human life. To this day, at least in Italy, no institution is truly able to protect the inhabitants of the plains from the effects of an unwise use of mountain soils. What is even worse is that the overuse of the territory has now expanded from the mountains and the hills to the plains, due to the cementing of riverbeds, which prevents the dispersion of water, causing traumatic consequences for urban settlements.

The Mediterranean, especially in the north, is a sea surrounded by mountains. These mountains are tall and wide, spanning the Spanish Pyrenees, the Alps, the Apennines, and the mountainous systems of the Balkans, up through Anatolia and the Caucasus. They are compact and their valleys are narrow. Passes between countries are rare, and until the late modern age, difficult to cross. It must be remembered that the Mediterranean area, besides the many towns on the coasts and the hills, where ancient civilizations and cultivations flourished, is a difficult territory with a skyward vertex, as noted by Braudel (Braudel, 1972), characterized by cold and snowy winters just a few kilometres from the coast. An example of this is Mt. Etna (3263 m), in Sicily. At sea level there are citrus and banana cultivations. Further up vineyards and olive trees can be found, then forests with oak trees and finally alpine vegetation covered by snow for nine months of the year. A similar example in Spain is the coexistence of the snowy mountain tops of the Sierra Nevada close to the Mediterranean coast, and Andalusia, rich in population, with its highly cultivated plains and foothill areas, yet semi-deserted and wild in the mountain areas.

For centuries thousands of people, alone or accompanied by their flocks, have temporarily migrated from the mountains to the valleys. The mountain men have constituted a large part of the workforce that cultivated plains infested by malaria, avoided by those who had a better way to earn a living. Mountaineers travelled throughout Europe to perform hard jobs: they built roads and canals, worked as bricklayers and artisans anywhere they could. The economic activity of the Mediterranean valleys and hills, as well as that on the other side of the Alps and the Pyrenees, cannot be fully understood if we do not take into account the contribution of the mountain inhabitants.

One of the most complex situations in the Mediterranean is that of the hills. These are areas of ancient settlements and that have been transformed by human labour. Among the hills highly transformed in this manner are those of ancient Catalonia, Languedoc and Provence; in Italy those of Piedmont, Veneto, and central Italy. These will be the privileged areas in our discussion. It is an area that, between the Middle Ages and the twentieth century, has been characterized by a series of accurate and labour intensive land settlements, which we will later discuss.

Finally, we come to the plains. In the Mediterranean area, for the entire period taken into consideration, they enjoyed only a potential prosperity because of their agricultural activity. Their main problem was the imperfect control of rivers and streams that periodically flooded the land with disastrous effects. Floods became more and more frequent as deforestation progressed into valleys and hills, caused by the need to obtain more wood or more arable land. An even worse threat to human settlements was the transformation of rivers into swamps and the consequent diffusion of malaria. Attempts at land reclamation began with medieval tillages and continued until the twentieth century, when it became possible to drain land through the use of water-scooping machines.

The Impact of Man

Between the Middle Ages and the modern era, the northern part of the Mediterranean was the most densely populated in Europe. Many people crowded the small amount of infertile land, concentrating both in cities and in the surrounding countryside.

Since the Middle Ages cities had a strong presence, especially in Italy. Until the seventeenth century Italian cities were among the largest in Europe and certainly among the most densely populated areas in Italy. Here are some significant statistics: in the sixteenth century the population of Milan and Venice was about 100,000 inhabitants, Genoa had 58,000, Florence, Palermo and Rome, 55,000 (Malanima, 1995), whereas many other Italian cities held between 20,000 and 50,000 inhabitants. In medieval times, the only regions that had a development comparable to that of Italy were Flanders and Brabant in northern Europe, although

hgl pop density

their cities were smaller in population. The density per square kilometre of central and northern Italian cities remained among the highest in Europe during the entire modern era.

'Dominant' cities became a prevailing environmental factor, since their power allowed them to spread through countryside and forests. These were manufacturing cities that needed to feed a population not employed in agriculture. They needed therefore a constant surplus of cereals, but also wine, oil, meat and raw materials for their manufactures.

The effects of this demographic density were the following. A precocious and widespread anthropization of the environment, whose effects are still evident to this day. As we know, artificial transformations of natural landscapes cause damage to the ecosystem, including desertification, erosion of the soil and pollution, all phenomena that intensify with the increasing need to procure food for a growing population. A second consequence of the demographic density was the development of labour intensive agriculture, both because of the relative costs of the production factors (land was the scarcest and most costly labour production factor) and because of the reduced possibility of raising working animals, since they competed for spaces otherwise dedicated for growing foods for human consumption.

The average human energetic efficiency is similar to that of a horse, amounting to a figure of about 13 percent. However, even though the level of useful power efficiency might be similar, the absolute performances are not. Human beings cannot sustain levels of useful power higher than 70-100 Watt, while a draught animal can work for hours at levels between 500 and 800 Watt.

One hour of an oxen's work can substitute 4 hours of human labour; that of a horse 5 to 8 hours (Wrigley, 1988). However, when the energetic balance of the two animals is considered, there are also other factors to take into account. Bovines provide milk and require less maintenance than a horse, in addition they are ruminant animals and therefore are only a partial burden to humans. Horses' needs, on the other hand, are more specific. Horses require the presence of a field sown with oats, which subtracts space from the cultivation of cereals for human consumption, such as wheat and rye (Malanima, 1995). Therefore, due to the scarcity of tillable land, many farmers in the Mediterranean area preferred oxen to horses as draught animals in the fields and for transportation, even if they were less efficient. In the hills of central Italy, farmers who had only one or two working animals, frequently used cows instead of oxen, because cows had the advantage of providing milk for the children and to make calves.

The scarcity of working animals remained a handicap for Mediterranean agriculture up until the mid-twentieth century. In fact it had as consequence a scarcity of animal energy to use in the fields and for transportation, as well as a scarcity of manure to be used as fertilizer in agriculture.

Another consequence of demographic density was the early anthropization of the forest, which became assimilated within human settlements. The *Sylva* was integrated into the ecosystem (Delort-Walter, 2001). Human beings colonized the forest to get extra food for themselves (see the case of the domesticated chestnut tree) or for their animals (like in the case of acorns used to feed pigs). In addition, the energy generated by firewood was necessary for human activities, both in agriculture and manufacturing. Wood was essential for making tools as well as dwellings and for transportation; however, with every sizeable demographic growth and the need to devote more land to farming, the forests receded, which was a serious environmental risk.

emphass an env. change as risk / degradation / rather than human progress

Historical Evolution

Mediterranean societies sought to create agrarian systems capable of meeting the demands of people and of the other sectors of the economy without exhausting the soil, since this was the most important collective resource.

Some widespread agrarian systems, especially those that integrated in the same fields agriculture and cattle raising, worked at a low entropy level. As a matter of fact, much of the waste was reused (manure, human excrements, leftovers, ashes) for new productive cycles.

The development of labour intensive agriculture was also relatively environmentally friendly. The aim was to safeguard and restore soil fertility while drawing the greatest possible returns. This was achieved by systems of slope cultivation that held on to the soil, preventing erosion (through stepping and terracing); rotations maintaining the fertility of the land; plantings requiring little water, including trees; water protection strategies in the valleys; raising and use of the appropriate animals (oxen, pigs, sheep).

The 'Malthusian Centuries': Growth, Crisis, and Recovery of the Mediterranean Population and its Consequences on the Environmental and Agrarian System

When looking at population figures in the Middle Ages, we notice that from 1000 to 1300 there was substantial demographic growth in the Mediterranean area. From the twelfth century to the first half of the fourteenth century, the demographic growth of the Middle Ages is characterized, in the northern Mediterranean, by tillage, an increase in the cultivation of domesticated trees, as well as an increase in hydraulic improvements on the plains (see Figure 3.1).

A major role in this process was played by the cities, which in the northern and central part of Italy were organized in *communes* (city states). In southern France similar forms of government started to appear in cities such as Lyons and Toulouse, which organized their own rural areas around the city. The capital circulating in cities was used to acquire more land and, partly, also to till, build houses and create roads. This was an era of heavy tillage and deforestation. The 'creation of the Mediterranean landscape' had begun, characterized by sparse settlements and family farms. The work of families who devoted themselves entirely to the cultivation of the land became an essential element of agricultural production (see Figure 3.2).

city – country relatshp

**Figure 3.1 Star-shaped field arrangements in reclaimed swampland at
Enserune in Hérault, France**

    In this era domesticated tree cultivation expanded considerably in the
Mediterranean region. One of the most important was the vine. Until the High
Middle Ages vines were cultivated in enclosures near cities or villages. In areas
where a landowner would give a worker the permission to cultivate his land with
various sharecropping contracts, vine cultivation and the sharing of grapes were
usually excluded from the contract. Vines are plants that require much care.
Usually they were cultivated in small closed plots with narrow rows, on a dry stake
or as a bush.

Figure 3.2 A. Lorenzetti, 'The Effects of Good Government in the Countryside', Siena, Palazzo Pubblico (1338-40). Courtesy of Comune di Siena

In the late Middle Ages (from the second half of the thirteenth century) the vine began to emerge from enclosures, made its appearance in the arable fields and became part of the Mediterranean landscape, from Catalonia to Provence, to Central Italy. Quite often, it was planted without the intervention of the landowner. As a matter of fact, throughout various eras, vine planting was at the core of many long-lasting emphyteusis contracts that were common until the twentieth century (like the Catalonian *rabassa morta*). However in other cases, like in the Italian *mezzadria* (sharecropping on a 50/50 basis) the landowner had to plant vines and olive trees at his own expense. The sharecropper was the one who maintained them, but in this case the agrarian contract was much shorter than the emphyteutic. Besides, every year free of charge, sharecroppers had to dig for the landowner a certain amount of ditches and maintain the rows in good condition and without empty spaces.

On the plains, vines were grown in connection with trees at the edge of the fields, close to ditches that had been created to allow water to run off. In the hills, the slow reshaping of the relief had begun. Attention was focused on soil protection, to avoid the loss of important nutrients and to prevent the process of desertification. Deforestation of the hills required immediate intervention in defense of the arable land in order to maintain its fertility, and this was usually accomplished through the control of surface waters.

Effects of the Plague of 1348

The consequences were not always positive (see Figure 3.3). There were negative environmental effects in terms of soil washing and soil erosion. Due to changes in the eating habits during the great demographic expansion of the Middle Ages, there was a decline in the use of pork products and new types of condiments were introduced, such as walnut oil and olive oil (especially in southern Italy and Spain, in the Seville area) to replace the usual lard and suet.

The period 1347-80 was characterized by demographic crisis. The steep population decline was accompanied by a complete unravelling of economic organization and extremely low agricultural production. The effects of the demographic crisis of the XIV century included the regrowth of forests, the expansion of uncultivated land, and an increase in the number of animals. In addition, more land became available for those who had survived the plague. Much of the land was now unfarmed. Hunting and sylvopastoral activities, not yet integrated into the agricultural system, regained importance. Pastures also became more widespread because they required less manpower.

Figure 3.3 A. Lorenzetti, 'The Effects of Bad Government in the Countryside', Siena, Palazzo Pubblico (1338-40). Courtesy of Comune di Siena

The old houses in the abandoned villages disappeared along with their gardens and orchards. The number of vines declined as there were no longer enough men to take care of them. Wherever they still remained, they were no longer farmed by daily workers because salaries had increased, but rather under lease contracts or, particularly in Italy, under sharecropping contracts. The work of the sharecropper, for those owners who chose short-term contracts, was the least expensive and at the same time the most reliable, in this as in other arboreal cultures.

In Aix-en Provence, where many villages had disappeared, between the end of the fourteenth century and the first decades of the fifteenth century, the vineyard almost vanished. In Languedoc there was a true catastrophe as a consequence of the Black Plague and the Hundred Years' War. It would be enough to take a look at the data mentioned by LeRoy Ladurie (*Les Paysans de Languedoc*): in one village where in 1353 there had been 40 vineyards, only 6 remained by the end of the fifteenth century. The decline of the vine was definitive wherever the climatic conditions were not favourable, as in parts of France.

From the end of the fourteenth to the beginning of the fifteenth century, though there were many fewer people, both land and capitals remained intact. There was an increase in salaries, more abundant resources for the survivors, a decrease in the price of cereals due to diminished demand and more land becoming available for the producers.

15c-16c Recovery

During the fifteenth- to sixteenth-century recovery, growth was not uniform and population density remained different in different areas. In the sixteenth century, Italy, like the Netherlands, had a density of 40 inhabitants per square kilometre, but 100 in the Milanese territory; France had an average of 34, England slightly less; Spain and Portugal 17 (but the Kingdom of Valencia 30), while there were only 2 inhabitants/km^2 in Russia and in the Scandinavian countries. The population density in the Mediterranean area affected the prices of wheat that were the highest in Europe till the end of the sixteenth century.

In the Mediterranean area at the end of the sixteenth century, 60 million people were unequally distributed. Between 1500 and 1600 the population of the Mediterranean doubled. Population growth was very intense, particularly between 1450 and 1550, and then it slowed down. Let us examine the case of Sicily: in 1501 the island had a little more than 600,000 inhabitants, whereas in 1583 it had over 1 million. The demographic growth in turn determined the price revolution, which was mostly its consequence. People living in areas that had not been so badly hit by the plague moved towards the Mediterranean coasts. There was a migration from the mountains and the highlands to the abandoned plains and valleys. Thus, Catalonia was repopulated by immigrants from the French Roussillon; the Italian Adriatic coast by people coming from northern Italy, but also by Slavs and Albanians who arrived from the opposite shore.

In the sixteenth century, demographic growth pushed the population of the Mediterranean areas once again up the hills but also down into the plains, where they had to face and fight malaria, which threatened their survival. The reward in the case of victory was the conquest of more fertile land, a mirage for the starving populations of the Ancient Regime.

Land drainage projects were long and expensive, and no owner, no matter how big, could afford to undertake them alone, also because hydraulic problems concern always vast territories and require the investments of huge capitals, which only great commercial cities, like Venice, Milan and Florence could undertake. The operations, before the era of the water-scooping machine, were undertaken with the slow, although less costly, method of 'filling up'. Land drainage projects were also undertaken by great capitalists in lower Languedoc (the work continued from the end of the sixteenth century to the end of the seventeenth), and in coastal Spain.

The Conquering of the Plains Represents the Greatest Success of the Mediterranean Agriculture

In central Italy, *poderi* (family farms) and sharecropping became the key to reclaiming uncultivated land and turn woods and bramble into arable fields.

In the Marche region, the agricultural recovery relied on emphyteutic tenants, with grants *ad pastinandum*, at the end of which the land, drained and ready to produce, was divided between the owner and the tenant.

Wherever population grew, agriculture grew accordingly. Once again the number of trees increased. Vines were once again on the rise. In the Mediterranean area, between the sixteenth and nineteenth centuries, they were rarely cultivated alone. More frequently, like in the late Middle Ages, they were planted in the same fields where cereals and olive trees grew. The landscape, at this point assumed very precise and regular characteristics. Vines were grown in rows on hillsides, on trees, or at both sides of ditches in the plains. The English philosopher John Locke described them in 1667: 'From Bordeaux to Cadillac … in many places the vineyeards set thus: 2 rows of vines and between them 3 or 4 times their breadth of ploud land for corne' (Locke, 1953, p. 239). Locke is referring to the area around Bordeaux, but this system was practised also in Provence. It continued until the nineteenth century, when agronomists criticized it, since ploughs damaged the roots of the vines. The Italian 'piantata' (planting) or 'la plantade', which began in the second half of the sixteenth century, and that LeRoy Ladurie described for the Cevennes, satisfied many agricultural and economic needs. The vines high on the trees were less susceptible to the humidity of the plains, and the supporting trees furnished 'forages' for the livestock. Thus, from the same land, one could get three or four products (cereals, wine, oil, forages).

It was during this time that the most typical features of the Mediterranean landscape began to emerge. The first was the arrangement of the hill soil through terracing and grass edging, and of the plains through the thick network of drainage ditches. In the plains, alongside the creation of the macro hydraulic network entrusted to the state magistrate, farmers developed a complementary micro hydraulic network, connected to the first one, which they maintained themselves guided by their deep practical knowledge.

The second feature was the presence of trees in the arable land. Their number had been increasing steadily since the High Middle Ages: not only vines, but also olive trees, walnut trees, citrus trees, mulberry trees and almond trees. Chestnut trees spread in the high portions of hills and mountains. They were known as the 'trees of bread', because they ensured the survival of the populations of Vivarais, of the mountains of the Cevennes, of all the Italian Apennines, of Corsica, of the Greek and Macedonian mountains. In some areas their cultivation was a one-crop economy.

The cultivation of some plants used in the manufactures expanded. They were plants that needed labour intensive farming, like hemp in the Po valley and *pastel* in the countryside around Toulouse. There was also a big growth in the number of mulberry trees, from Spain to Italy. The mulberry tree required, like hemp or vines, a large amount of work. The peasants families of these areas were engaged in farming – both working the land and raising livestock – all year around, including winter.

The amount of agricultural products increased mostly due to the expansion of the cultivated land. The yield of cereals, however, remained low, due to the scarcity of livestock and thus of manure. In fact deforestation progressed, thus reducing the space devoted to woods and uncultivated land. As a consequence, hunting and wild fruit harvesting decreased. Hunting became an occupation for the upper classes, whereas before it had been open to all.

The construction of the 'anthropised' artificial landscape resumed. The goal of the hill arrangement was 'to ensure the defence of the agrarian soil, a more balanced hydraulic economy and a more efficient land utilization, by constructing horizontal fields of appropriate width' (Sereni, 1961). The arrangement of hillsides with grass planted in contour strips and terraces was not new. It was done during periods of demographic growth in those territories where people had been residing since antiquity and where agriculture was the only possible activity.

In the Mediterranean landscape between the sixteenth and the nineteenth centuries a variety of methods was employed for the mountainside and hillside arrangement. One consisted of grass strips planted in contour edges, another employed 'lunettes'(circles of stones and dry twigs arranged around every single tree or around two or three, so that the small amount of soil would not be swept away); another consisted of steps without real terracing; and finally there was terracing. In this case the terraces were supported by embankments made with the stones dug from the tilled soil (the same stones that were used to build the small peasant shelters in the areas far away from villages, or the houses of the tenants who lived permanently on the fields).

The care taken by humans to protect the hill and mountain soil, however, did not prevent environmental degradation, which was evident in many parts of the Mediterranean. In particular, deforestation and the farming of steeply sloping areas had very negative environmental effects. Abbé Rozier (1785) complained about the ill effects of cultivating terraines along steep mountainsides and the almost irreparable mistake of cutting the forests covering the mountain tops, leaving only bare rock where trees could not be replanted, and allowing silt to reach the plains below.

Some sovereigns in the sixteenth century tried to put a stop to deforestation. In France, Francis I and Henry II's love of hunting had as a consequence the issuing of decrees regulating the management of forests in the first decades of the sixteenth century. In 1669 Colbert issued *l'ordonnance*, aimed at a more rationalized exploitation and protection of royal, city and church-owned forests. In Tuscany in 1559, Cosimo I, concerned about the repeated flooding of the river Arno, sought to remedy the situation by prohibiting the cutting of forests within the first half mile of the highest peaks of the Apennines; in 1564 this distance was increased to one mile. These were followed by other restrictive measures, including those against cutting more than ten chestnut trees without permission from the government, against cutting any ashes, pines or elms, and against cutting deciduous trees under fifteen years old (Vecchio 1974). This policy, concerned less about the environment than about the safety of the land and the rational exploitation of forest resources, was abandoned in the second half of the eighteenth century. Then, a policy based on physiocratic theories and favourable to the creation of bourgeois private property was applied to forests. Owners were now free to use their forest resources as they leased, without any interference from the government. By the nineteenth century, this had caused an ever worsening erosion of mountain and hillside soil and increasing problems in the hydraulic situation of the valleys and plains.

The Heyday of the Man-Made Countryside

In the late modern age (eighteenth and nineteenth centuries), the Mediterranean area was still characterized by demographic growth, although less so than other European regions. During this period the 'man-made countryside' and peasant labour reached their heyday. The land very rarely belonged to those who worked it. Landed property was in the hands of old noble families, of the church, or it was acquired by families belonging to the manufacturing or commercial sector or to the professions.

Peasant families were often part of an estate, in which each member of he family was required by contract to work. According to sharecropping contracts in Catalonia, France and Italy, the labour force represented by the family was required to work for the estate all year long, as a look at the Tuscan calendar can show. In these contracts the labour force supplied more labour power than in any other type of contemporary agrarian contract.

We can find examples of this in the bookkeeping of Tuscan estates, the *fattorie*, comparing their data with the data from parish registers. Around the middle of the nineteenth century, at *La Cava*, one such *fattoria* on the low hills around Pisa, there were sixteen *poderi*, i.e. family farms, farmed by sharecroppers. The density for every hectare of land was of 1,4 sharecropper family memebers, and an adult worker for every two hectares of land (almost all of which was cultivated). The draught animals were generally two oxen in each *podere*. Here the number of workers was higher than in other less cultivated areas of Tuscany. A comparison with the sharecropping numbers of less populous France makes the density in Tuscany stand out even more (Tourdonnet, 1979-80). As I have already

Why is this chapter in the book?

mentioned above, these families were obliged by contract to work all year on the farm, which absorbed all their labour force. *however — waterterrace*

These sharecroppers, the day labourers who helped them (paid by the owner), and the owners of small lots, accomplished the organization of fields on hillsides and plains. On hillsides, the energy of men and women was the only force that built kilometres of terracing, edging, and other arrangements that are still visible today in various areas of Italy and the Balkans. We could still see similar ones in many other areas, except that they are now buried under the wild vegetation that has reclaimed the land after the last fifty years of neglect. The unique landscape of the Cinque Terre in Liguria, offers a good example of the importance of human work in shaping nature, in creating and saving soil for agricultural purposes and in cultivating vines in a very traditional way up to this day. Here women's work has been particularly important, as the men were mostly sailors.

The terraces are often masterpieces of technique in the use of stone and the channeling of rainwater. Soil was often carried up by hand, from the lower levels or from the valleys, to create the artificial terraces. These surfaces were then surrounded by 'dry' stone walls with water drainage ditches at the base. We have calculated that in the nineteenth century in the Chianti region, 100 metres of terracing, built according to the best criteria, would cost the equivalent of 250-500 days of labour, with peasant labour calculated on the basis of a daily wage. This amount, multiplied by the hundreds of kilometres of artificial terraces created in the Mediterranean area between the Middle Ages and the twentieth century, gives some idea of the incredible amount of work, almost all human (with minor help from animals for the transportation of the stones, wherever possible), involved in creating these artificially level surfaces on hillsides, and some idea of the costs in economic terms of such an operation, possible only in centuries with extremely low labour costs. Let us try to quantify the phenomenon for another Italian area, the Cinque Terre, in Liguria. Here, from medieval to modern times, the portion of terraced land reached a maximum of about 1,400 hectares. In each cultivated hectare there where 20 to 25 bands with wall lengths between 2,500 and 2,000 metres, that is, a minimum of 56 million metres of dry stone walls. Adopting the Chianti calculation, this means something like a minimum of 140 million days of work, a full time job for 2,000 people for nearly two centuries.

The most perfect hill arrangements are perhaps found in Tuscany, where herring-bone cultivations had been practised at least since the eighteenth century, and where in the nineteenth century century the so-called *colmate di monte* originated, and spread later to other parts of central Italy.

This hill arrangement, invented by an administrator of the marquis Ridolfi named Agostino Testaferrata, aimed not only at protecting the soil from rainwater, but also at using silt driven by rainwater to improve hillside soil for agricultural purposes. Testaferrata implemented a method that consisted in digging cavities in one part of the hill by building an embankment. From these cavities various ditches originated and ran down the hillside. When rainwater had filled the cavity, the embankment was removed and the waters fell into the ditches, which contained soil shovelled from the sides. This soil, together with silt that the water collected on its downward journey, went to fill the lower cavities that needed to be filled, and which had previously been embanked. Thus, step by step, the system of filling up the mountain allowed the creation of new regular fields along the hillside, on slopes that were gradual and easy to cultivate. The hill drainage was complete, however, only with the creation of drain ditches that could prevent damage caused by erosion and landslides. It was therefore necessary to trace a system of drainage ditches allowing the water to descend gradually to the plain. The ditches had to be pitched just enough to allow the flow of water. The water began at the main ditch at the top and flowed down to the following one, parallel to the first. Upon hillsides prepared in this manner, the characteristic 'herring bone' cultivation was imposed, which is still considered the most perfect example of hillside agrarian landscape architecture.

The End of a World

In all the countries affected by industrial development, including Italy, the second half of the twentieth century saw the abandonment of the countryside and the end of traditional peasant labour devoted to the construction and protection of the soil.

When agriculture was no longer the primary activity for the survival of the population, and the soil ceased to be the 'great mother' of everyone, its safeguard ceased to be economically necessary. The consequences are before our eyes. Over the last forty years the degradation of the Mediterranean landscape and environment have been so rapid that it is visible in images of the same place across a short span of time. Terraces threaten to collapse or have already vanished; hill soil has washed away because of the end of water control due to the disappearance of ditches, a consequence of the use of tractors, meanwhile ditches and wind breaking hedges have vanished from the plains, soils and riverbeds have been cemented. No doubt, labour and human energy have their price; and one can not think of safeguarding everything that these have created in the past centuries when they were economically the cheapest production factors. Nevertheless, all the progress of contemporary scientific research has not yet found an adequate substitute for them.

References

Braudel, F. (1972), *The Mediterranean and the Mediterranean World in the Age of Philip II*, 2 vols., Harper and Row, New York.

Casavecchia, A. and E. Salvatori (2001), *Il parco dell'uomo. 1. Storia di un paesaggio*, Parco nazionale delle Cinque Terre, Riomaggiore.

Delort, R. and F. Walter (2001), *Histoire de l'environnement européen*, Presses Universitaires Françaises, Paris.

Locke, J. (1953), *Locke's Travels in France 1675-1679, as Related in his Journals, Correspondence and Other Papers*, ed. with intro. and notes by J. Lough, Cambridge University Press, Cambridge.

Malnamia, P. (1995), *Economia preindustriale. Mille anni: dal IX al XVIII secolo*, Bruno Mondadori, Milano.

Rozier (1785), *Cours complet d'agriculture*, 12 vols., vol. IV, Paris.

Sereni, E. (1961, 1997), *History of the Italian Agricultural Landscape*, tr. R.B. Litchfield, Princeton University Press, Princeton.

Tourdonnet, A. de (1879-80), *Situation du métayage en France*, impr. de la Société de typographie, Paris.

Vecchio, B. (1974), *Il bosco negli scrittori italiani del Settecento e dell'età napoleonica*, Einaudi, Torino.

Wrigley, E.A. (1988), *Continuity, Chance and Change: the Character of the Industrial Revolution in England*, Cambridge University Press, Cambridge.

PART 2:
ENERGY AND POLITICS

Part 2 , on Energy + Politics, includes chapters on

PART 2
ENERGY AND POLITICS

Chapter 4

Energy and Sustainable Development

Jürgen-Friedrich Hake and Regina Eich

Systems Analysts, Germany

The term 'Sustainable Development' plays a major role in the debate on social development extending into the future. It aims at expressing the concept of the need to make constant economic growth coexist with the preservation of the environment, while at the same time maintaining social justice. The starting point for this debate is the statement that, although social development in the past has *Inequalities* brought prosperity for many people, a much greater proportion of mankind still lives in poverty in substandard conditions. Closely related to this status description is the view that both the nature of present-day economic activities and the current patterns of behaviour and consumption in the industrialized countries are not suitable as a model for future development, and actually might have serious repercussions on ecological, economic and social subsystems. The concept of sustainable development thus links the issue of conserving the natural basis for life for future generations to the desire for economic prosperity and social development of the people living in the present. — *3 pillars of S.D.*

Providing useful energy for eight to twelve billion people in the future, of which a large percentage will probably live in conurbations with several million inhabitants each, has several geopolitical dimensions, which should not be ignored. The unequal geographical distribution of global energy reserves and the demand, which strongly deviates from this distribution, require a flexible and also robust system of international trade and supply relations. The commitments to active climate protection resulting from the United Nations Framework Convention on Climate Change and from the Kyoto process can also only be fulfilled in close international cooperation. The leeway for shaping national policies decreases – at least at first sight – as this network of global policies becomes more concrete and binding.

increasing internationalisation of energy policy + loss of control at the national level ?

Sustainable Development

Origin and Background of the Term

In 1972 an 'Action Plan for the Human Environment' was adopted by the UN General Assembly. It contained measures and agreements in the following fields: the worldwide acquisition of environmental data, the sector of environmental research, the exchange of information, the careful handling of resources, the creation of global environmental administrations and the targeted education, training, and information of the population. The first step was to establish an independent environment secretariat designated 'UNEP' with headquarters in the Kenyan capital Nairobi. In this context, Maurice Strong, the first Executive Director of UNEP, coined the term 'ecodevelopment' (which was replaced in the early 80s by the term 'Sustainable Development') for this new development strategy. The aim of the programme established under the control of UNEP was to design a middle road between the conflicts emerging in the early to mid-70s between the eco- and techno-centric positions concerning the value of the environmental protection concept. The concept of ecodevelopment, which in the beginning was mainly meant as a development approach towards supporting the efforts on behalf of the predominantly rural regions in Africa, Asia, and South and Latin America, contained basic theoretical assumptions suggesting a new definition of development, growth and prosperity going beyond the original target groups in the Third World.

In the history of the origin and further development of the overall concept of Sustainable Development, the report, Our Common Future', presented in 1987 by the United Nations Commission on Environment and Development, represents a milestone despite the work already performed by UNEP before. The commission is generally called the 'Brundtland Commission' after its chairwoman, the Norwegian politician Gro Harlem Brundtland. Starting with the problem of finite resources, on the one hand, and the unequal distribution of prosperity and resource consumption, on the other, the Commission formulated the following definition of Sustainable Development: 'Sustainable Development is development that meets the needs of the present without compromising the ability of future generations to meet their own needs' (World Commission on Environment and Development: Our Common Future, 1987). Sustainable Development thus centres on a fiduciary use and at the same time on the conservation of the basis for life available to humans for future generations. In its report, the Commission also referred to the necessity of limiting the inputs of anthropogenic material into existing ecological cycles. Relative to the greenhouse effect, the proposal was to restrict the energy-related emission of greenhouse gases against the background of global climate change.

Dimensions of Sustainable Development

The report of the Brundtland Commission contained a multidimensional deployment of the overall concept of 'Sustainable Development'. Today, two approaches can be distinguished. The three-dimensional model – also called the 'three-pillar model' – is based on the assumption that the concept of Sustainable Development rests on the three dimensions of industry, society and the environment. In contrast, the four-pillar model – as well as Agenda 21 – involves a social, ecological, economic and (political-) institutional dimension (Federal Ministry for the Environment, Nature Conservation and Nuclear Safety, 1992). Both models are currently being applied. The essential difference is the importance of institutional aspects. The supporters of the three-pillar model understand the political-institutional level as the 'framework' needed for the implementation of

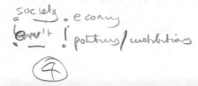

the overall concept of Sustainable Development. The advocates of the four-dimensional approach have recently changed their position, no longer regarding this aspect as a mere boundary condition, but rather assigning it the status of an independent dimension.

So far, there is no agreement on which approach should be pursued for formulating national sustainability strategies. Thus, the Federal Republic of Germany, as a signatory state of Agenda 21, has also supported the four-dimensional approach. However, the final report of the study commission 'Protection of Man and the Environment' presented in 1997 supports a three-dimensional sustainability concept.

Indicators of Sustainable Development

Very abstract normative goals and value concepts must be transformed into concrete 'instructions for action'. This requires indicators allowing development to be characterized as comprehensively as possible. The task of indicators is to provide information about the state of the current system. With the aid of indicators, it should be possible to tell whether the state of the observed system is felt to be satisfactory or if changes must be made. Furthermore, indicators should be quantifiable as far as possible. Time series are required for an assessment of the dynamics of the system.

As a general rule, the processes taking place in connection with Sustainable Development always contain social, economic and ecological aspects with different weighting. Individual indicators examined in isolation prove to be less effective here. Rather, indicator systems capable of reflecting both the current state of the system considered as a whole and of showing current and future developments (so-called 'trends') should be conceived. Since the appearance of the Brundtland report, numerous national and also supranational actors have developed a set of indicator systems so varied as to be unmanageable. The indicator systems of the international organizations very often relate only to a partial aspect of Sustainable Development, e.g. to an energy supply oriented towards the overall concept of Sustainable Development. The danger here is that the connection to the superstructure is lost or completely missing.

Indicators of the United Nations Commission on Sustainable Development

The indicator system of the United Nations Commission on Sustainable Development (CSD) is oriented to all 40 chapters of Agenda 21. Based on the four dimensions of Sustainable Development, the economic, ecological, social and institutional aspects are reflected in the CSD scheme by a set of indicators. The CSD starts out with a set of 138 indicators, including 23 that can be allocated to the economic, 38 to the ecological, and 15 to the institutional dimension (Federal Ministry for the Environment, Nature Conservation and Nuclear Safety, 2000). At present, the set of indicators developed by the CSD is being applied and tested for practical suitability in a kind of testing phase in 20 selected countries. Germany is among the countries participating in this 'test run'. Table 4.1 gives a survey on the structure and the most important topics of the CSD indicator set.

Table 4.1 CSD key elements on sustainable development

Economic Indicators	Environmental Indicators	Social Indicators	Institutional Indicators
economic performance	climate change	poverty	strategic implementation of SD
trade	ozon layer depletion	gender equality	international cooperation
financial status	air quality	nutritional status	information access
material consumption	agriculture	mortality	communication infrastructure
energy use	forests	sanitation	science and technology
waste generation and management	desertification	drinking water	disaster prepardness and response
transportation	urbanization	healthcare delivery	
	coastal zones	education level	
	fisheries	literacy	
	water quantity	living conditions	
	water quality	population change	
	ecosystem		
	species		

Indicators of the OECD

The *Organisation for Economic Co-Operation and Development (OECD)* first developed a catalogue of environmental indicators in 1993. On this basis, five years later, it adopted an indicator list, in which the 50 socio-economic indicators were also included in addition to the environmental indicators previously derived. These 50 indicators mainly relate to the sectors of industry, population, unemployment, consumption expenditure (by national and private households), energy, transportation and agriculture. Table 4.2 gives an outline of the structure and topics of the OECD indicator set. Moreover, it demonstrates how the energy section is embedded within the complete indicator set.

During the world economy summits of 1989 and 1990, the G7 states agreed in the future to integrate economic and ecological issues more strongly into decision-making processes. For this purpose, the OECD was requested to develop a set of environmental indicators (Radke 1998).

The environmental indicator set should serve to analyse the environmental achievement of individual countries. Environmental indicators can be used to analyse the environmental efficiency of a country, a region or an economical sector if these indicators are based on measurement and also identify the driving forces for change. The indicators can be connected with quantitative goals and qualitative goals, e.g. efficiency in human activities or foresight in the use of natural resources and development. In 1991 the OECD presented a first set of indicators connecting environmental and economic factors. The indicators can be classified as follows:

- Harmful materials and waste material entries into the environment;
- Emissions per unit GDP;
- Emissions per capita of the population;
- Temporal trends of GDP;
- Consumption of fossil fuels and pollutant emissions;
- Use profile of selected natural resources;
- Environmental condition;
- Defensive efforts of environmental protection; and
- Public perception of the environmental problems.

The OECD defines its concept as a 'set of core indicators'. The aim is to be able to construct interlinkages and interfaces between the individual indicators and, on this basis, between the indicators and the different policies. Furthermore, the OECD underlines that a set of indicators must not be used in isolation, but that additionally available information such as the respective ecological, geographical, social and economic context as well as the political and constitutional boundary conditions must be taken into account.

Table 4.2 The OECD set of 'core indicators'   *(includes IEA ?)*

Ecologic Indicators	Socio-economic Indicators
Climate Change	*GDP and Population*
CO_2 emission intensities	gross domestic product
greenhouse gases concentration	population growth and density
Ozone Layer Depletion	*Consumption*
ozone depleting substances	private consumption
stratospheric ozone	government consumption
Air Quality	*Energy*
air emission intensities	energy intensities
urban air quality	energy mix
	energy prices
Waste	*Transport*
river quality	road traffic and vehicles intensities
waste water treatment	road infrastructure densities
	road fuel prices and taxes
Water Resources	*Agriculture*
intensity of use of water resources	intensity of use of nitrogen and phosphate fertilisers
public water supply and price	livestock densities
	intensity of use of pestcides
Forest Resources	*Expenditure*
intensity of use of forest resources	pollution abatement and control expenditure
forest and woodlands	official development assistance
Fish Resources	
fish catches and consumption: national	
fish catches and consumption: global and regional	
Biodiversity	
threatened species	
protected areas	

apple of wheels

The OECD hopes that with the aid of its indicator list, which contains a category by category comparison of individual policies, it will be possible to turn international agreements – especially in the environmental sector – into concrete requirements and recommendations for action. Moreover, it believes that its set of indicators will provide its member states with a guideline for the development of national goals in this range of topics. In addition, the OECD advocates an international standardization of the respective indicator sets (OECD, Frameworks to Measure Sustainable Development, 2000).

Indicators of the World Bank

The set of indicators established by the World Bank is designated 'World Development Indicators'. It represents at the moment the most extensive set of indicators. Its advantage lies in the fact that the World Bank has already acquired data for about 150 states over a span of 40 years. The World Bank works with a total of 600 indicators, which can be assigned to the following six categories (Table 4.3): economic scope/economic structure ('World View'), data on population, labour market, income and property, poverty, health and education ('People'), environment ('Environment'), GDP, production structures, imports, exports, balances of payment, indebtedness, demand, supply, investments, money and prices ('Economy'), tax policy, infrastructure, science, technology, the state as an entrepreneur, stock markets ('States and Markets') as well as the scope and structure of trade with goods and services, cash flows and number of foreign employees ('Global Links') (World Bank: World Development Indicators, 2000).

Table 4.3 Indicator set of the World Bank

World View	People	Environment	Economy	States and Markets	Global Links
economic scope economic structure	data on population labour market income and property poverty health and education	environment	GDP production structures imports exports balances of payment indebtedness demand supply investment money money prices	tax policy infrastructure science technology state as an entrepreneur stock markets	scope / structure of trade with goods and services cash flows number of foreign employees

World Resources Institute

how are ecological foundations to the economy represented in the indicators.

An aim of the World Resources Institute is to observe and evaluate global development trends within the framework of its continuous 'World Resources' reporting scheme. This also includes roughly 20 economic and 80 social indicators for evaluation, in addition to ecological indicators. The economic parameters are concerned, e.g. with the scope and structure of the GDP, indebtedness, domestic and foreign investments, and energy, as well as price and goods indices. The social measurement quantities concentrate on the sectors of population, health, urban development, agriculture and nutrition (World Resources Institute, 2001).

Further Approaches

Apart from these supranational efforts, individual nation-states are developing their own indicator systems in response to the resolutions passed in Rio de Janeiro. Thus, in 1996, the US President's Council on Sustainable Development, instituted by Bill Clinton, presented their first set of roughly 50 indicators (SDI Group, 1996), which referred to ecological, social and economic matters (Table 4.4). As indicators for the social dimension, for example, the qualifications of school-leavers, internet access rate, income distribution or poverty rate were specified; economic measurement quantities included GNP, unemployment rate, per-capita savings, per-capita productivity, and also energy efficiency.

Table 4.4 Indicator set of the SDI-Group (USA)

Economic Indicators	Environmental Indicators	Social Indicators
capital assets	contaminants in biota	access to telecommunications
labor productivity	timber growth to removals balance	educational attainment by level
total materials per unit of investment and per PCE	metropolitan air quality non-attainment	life expectancy
investment in R&D as a percentage of GDP	status of stratospheric ozone	educational achievement rates
economy Management Index	greenhouse climate response index	percentage children living in poverty
consumption and government expenditures per capita	greenhouse gas emissions	people in census tracts with 40% poverty
home ownership	waste inventory	citizen's participation
percentage of households with housing problems	surface water quality	access to health care
vehicle ownership, fuel use and travel per capita	land use	homelessness
percentage of renewable energy	ratio of renewable water supply to withdrawals	population
	fisheries utilization	children living in families with at least one parent present
	invasive alien species	crime rates
	soil erosion rates	teacher training and applications of qualifications
	outdoor recreational activities	wealth distribution
	biodiversity	contributing time and money to charities

Furthermore, many European countries have meanwhile also adopted sustainability strategies with associated indicator systems. Among these are Sweden, the United Kingdom and The Netherlands.

Sustainable Development in the European Union

The concept of the European Union for Sustainable Development is strongly based on analytical work arising from concerns about environmental protection. Resulting from this ongoing process, topics of environmental concern have been identified and a set of indices has been specified (Table 4.5). Specific environmental protection goals have been formulated to guide development.

This scheme serves to orient not only the member countries of the Union, but all European states, and in particular the European Union entry candidates in Central and Eastern Europe.

Following the recommendations of the Commission for Sustainable Development (CSD), the statistical office of the European Union (Eurostat) developed a sustainability indicator set. Eurostat adapted the indicator catalogue of the CSD to the conditions of the EU countries and extended the list from 59 to 63 indicators. It is differentiated further between economic, social, ecological and institutional indicators.

The selection of indicators was determined by the following criteria. First of all data availability in the European Union was taken into consideration; next, availability of indicators, which supplement the present UNCSD indicator list and thereby refer to important European Union questions, which are not covered by the Agenda 21. Attention is paid to establishing an indicator catalogue consistent with the CSD list.

Table 4.5 EU indicators of environmental pressure and indices

Air Pollution	emissions of nitrogen oxides (NOx)	methane volatile organic compunds	emissions of sulphur dioxide (S02)	emissions of particles	and diesel o il by road vehicles	primary energy consumption
Climate Change	emissions of carbon dioxide (CO2)	emissions of methane (CH4)	emissions of nitrous oxide (N2O)	chlorofluorcarbons (CFCs)	emissions of nitrogen oxides (NOx)	emissions of sulphur oxides (SOx)
Loss of Biodiversity	damage and fragmentation	wetland loss trough drainage	area used for intensive arable	forests and landscapes by	and semi-natural forested areas	change in traditional land-use practice
Marine Environment and Coastal Zones	eutrophication	fishing pressure	development along shore	discharges of heavy metals	oil pollution at coast and at sea	halogenated organic compounds
Ozone Layer Depletion	bromofuorcarbons (halons)	chlorofluorcarbons (CFCs)	hydrchlorofluorcarbon s (HCFCs)	oxides (NOX) by aircraft	emissions by chlorinated carbons	emissions bof methyl bromide (CH3BR)
Resource Depletion	water consumption	energy use	permanently occupied by	nutrient balace of the soil	electricity production from fossil fuels	timber balance
Dispersion of Toxic Substances	pesticides by agriculture	persistent organic pollutants (POPs)	consumption of toxic chemicals	index of heavy metal emissions to water	index of heavy metal emissions to air	emissions of radioactive material
Urban Environmental Problems	energy consumption	non-recycled municipal waste	non-treated waste water	share of private car transport	people endangered by noise emissions	from natural to built-up area)
Waste	waste landfilled	waste incinerated	hazardous waste generated	municipal waste generated	during the lifetime of a number of products	waste recycled / material recovered
Water Pollution	(eutrophication equivalents)	ground water abstraction	hectare of utilised agriculture areas	used oer hectare of utilised agriculture	water treated / water colleted	emissions of organic matter as BOD

Table 4.6 Social dimension of SD in the EU indicator system

Eurostat SD Indicators			
Social Dimension of SD			
UN Main-Dimension	**Sub-Dimension**		**Eurostat Indicators**
		SOC 1	Population below the poverty level
		SOC 2	Parameter for the measurement of the income inequality
	Poverty	SOC 3	Unemployment rate
		SOC 4	Youth unemployment
Equality		SOC 5	Social development
	Sex equality	SOC 6	Comparison between woman's and men's salaries
	Child welfare	SOC 7	Child welfare
	Food	SOC 8	Food supply of the population
	Disease	SOC 9	Mortality due to certain diseases
	Mortality rate	SOC 10	Infant mortality
		SOC 11	Life expectancy at birth
Health	Sanitation	SOC 12	Population enrolled in the sanitary system
	Health Care	SOC 13	National expenditure for health
		SOC 14	Children's vaccination rate
Training	Level of education	SOC 15	Highest level of education
	Ability to read and write	SOC 16	Lowest level of education
Dwelling	Life circumstances	SOC 17	Space per capita
		SOC 18	Composition of the household
Safety	Criminality	SOC 19	Crime rate
		SOC 20	Population growth rate
Population	Change of the population	SOC 21	Population density
		SOC 22	Net migration rate

Source: Eurostat, 2001, 5

Social Dimension

The social dimension of Sustainable Development was described in 2000 by the UN according to the following indicators: equality, health, training, living, security and population. The UN divided these main topics still into subtopics. The EU modified and adapted the UN indicators in order to take into account the particular living conditions of the people in the European Union. Access to safe drinking water and contraception, for instance, were not considered particularly relevant.

Ecological Dimension

The EU act assumes that many human activities pose a threat to the environmental factors of air, land, water and diversity of species. For the usual reason of irrelevancy within the European context, Eurostat left out UN ecological indicators like: desertification; development of the coastal ranges and key ecological system areas. Instead, it added four new environmental indicators (industrial waste, waste treatment and refuse disposal, expenditures for environmental protection, and transportation of goods), and inserted these according to the UN system into the range economic indicators.

Table 4.7 Ecological dimension of SD in the EU indicator system

Eurostat SD-Indicators			
Ecological Dimension of SD			
UN Main Dimension	Sub-Dimension	Eurostat Indicators	
Atmosphere	Climate Change	ENV 1	Per capita missions of GHG
	Ozone layer	ENV 2	Consumption of the ozone layer damaging substances
	Air quality	ENV 3	Air pollution in cities
Land	Agriculture	ENV 4	Agriculture surfaces
		ENV 5	Nitrate pollution
		ENV 6	Use of pesticides
	Forest	ENV 7	Entire forest surface
		ENV 8	Wood impact relationship
	Urbanization	ENV 9	Utilization of land
Sea	Territorial waters	ENV 10	Acidification of territorial waters and sea
	Fishery	ENV 11	Fishing of endangered species
Fresh water	Water quantity	ENV 12	Intensity of water use
	Water Quality	ENV 13	BOD concentration in selected rivers
		ENV 14	Quality of water for domestic uses
Biodiversity	Ecosystem	ENV 15	Protected surfaces in % of total area
	Species	ENV 16	Number of endangered species

Source: Eurostat, 2001, 6

Economic Indicators

Economic indicators were selected in order to include the preservation of the standard of living for humans over time as one of the goals of Sustainable Development. Attaining this goal will involve implementing lasting production procedures and consumer habits. Sustainable Development means an efficient use of energy and material resources, effective waste management and sustainable transportation.

The EU selection of indicators follows to a large extent the UN defaults. However, some indicators were added, expressing the special conditions of the European Union national economies.

Table 4.8 Economic dimension of SD in the EU indicator system

Eurostat SD-Indicators			
Economic Dimension of SD			
UN Main Dimension	Sub-Dimension		Eurostat Indicators
Economic Structures	Economic efficiency	ECON 1	GDP per head
		ECON 2	Portion of the investments of GDP
		ECON 3	Portion of the individual sectors to GDP
	Trade	ECON 4	Inflation rate
		ECON 5	Deficit on the balance of payments on current account
		ECON 6	European Union and international markets
	Financial situation	ECON 7	Public debt
		ECON 8	Development assistance
Consumption and Production	Material consumption	ECON 9	Material consumption
	Energy consumption	ECON 10	Per head energy consumption
		ECON 11	REN
		ECON 12	Energy intensity of the individual sectors
	Waste	ECON 13	Increase of public waste
		ECON 14	Increase of industrial waste
		ECON 15	Increase of special refuse and special refuse disposal
		ECON 16	Radioactive waste
		ECON 17	Recycling of paper and glass
		ECON 18	Waste treatment and Refuse disposal mechanisms
	Transport	ECON 19	Increase of passenger traffic according to mode of transportation
		ECON 20	Increase in traffic of goods according to the mode of transportation
	Environmental protection	ECON 21	Expenditures for environmental protection

Source: Eurostat, 2001, 6

Table 4.9 Institutional dimension of SD in the EU indicator system

Eurostat SD -Iindicators			
Institutionelle Dimension of SD			
UN Main Dimension	Sub-Dimension		Eurostat Indicators
institutional frameworks	International Co-operation		
institutional capacities	Information accessibility	INST 1	Internet availability
	Communication infrastructure	INST 2	Communication infrastructure
	Research and Technology	INST 3	R+D Expenditures
	Disaster control	INST 4	Risks for human capital and nature capital

Source: Eurostat, 2001, S.8

Institutional Indicators

Sustainable Development cannot be achieved if the institutional basic conditions do not make possible the full participation of society in a national sustainable development strategy in compliance with international conventions.

Sustainable Development in Germany

The concept of Sustainable Development has met with approval in Germany, especially in the debate on the country's further internal development. In this context, a discussion and analysis of the development is taking place above all at the federal level. In addition, however, a multitude of regional and local initiatives towards implementing the overall concept of Sustainable Development have been undertaken at the individual state level (regional and local Agenda 21 processes). The following outstanding 'road marks' should be mentioned here in particular:

• The report of the Study Commission of the 12th German Bundestag (1990 to 1994) entitled 'Protection of man and the environment – evaluation criteria and prospects for environmentally compatible material cycles in industrialized societies';
• The environmental consultations of 1994, 1996 and 2000 of the Council of Experts for Environmental Issues;
• The study 'Sustainable Germany' of the Wuppertal Institute for Climate, Environment and Energy (1996);
• The Study Commission 'Sustainable Energy Supply under the Conditions of Globalization and Deregulation' of the 14th German Bundestag (1998-2002).

The first steps towards institutionalizing the overall concept of Sustainable Development were undertaken in Germany in the mid-90s. Under the title 'Steps towards sustainable, environmentally appropriate development', a discussion forum under the chairmanship of the German Federal Ministry for the Environment (BMU) was instituted in 1996 at the federal level.

Like the other states among the 180 or so which signed the Rio declaration, the Federal Republic of Germany committed itself to developing a national sustainability strategy in connection with the Agenda 21 process. Furthermore, in Rio de Janeiro the signatory states agreed to present their national strategies at the 2002 Conference in Johannesburg at the latest. Within the framework of negotiations on the formation of a coalition between the Social Democratic Party of Germany and Alliance 90/Greens, this commitment was taken up and included in the coalition agreement signed in autumn 1998. In line with this, during the summer of 2000, the Federal Cabinet adopted a bill according to which a Council for Sustainable Development was to be instituted by the Federal Government at the beginning of the following year. Its task was to participate in the development and formulation of a sustainability strategy for the Federal Republic of Germany. For this purpose, high-ranking representatives of different social groups in the Federal Republic of Germany – such as representatives of the two Christian churches, the consumer associations, the local authorities as well as industry and science – were appointed to this panel by the Federal Chancellor.

Moreover, the 'red-green' Federal Government agreed on the institution of a State Secretary Committee for Sustainable Development. In this committee, which is called the 'Green Cabinet' following the appointment of so-called 'Green Ministers' in the United Kingdom, the State Secretaries from ten of the total number of 14 federal ministries met. The Committee is chaired by the State Secretary of the Federal Chancellery. Its task also consists of developing a national sustainability strategy on behalf of the Federal Government.

The 'Green Cabinet' published its first considerations on Sustainable Development towards the end of 2001 in the form of a first draft of a national sustainability strategy for Germany. The 'Green Cabinet' selected a form of presentation composed of target levels with associated lower-level indicators. These four target levels include the topics of justice from one generation to the next ('Generation Justice'), quality of life, social bonds and international responsibility (Table 4.10). The target indicators and objectives developed by the State Secretary Committee will be presented in the following survey. The Committee stated furthermore that a plan of Sustainable Development was feasible only in the context of international cooperation. It is convinced, however, that Germany, with its national initiative, should furnish a major contribution to this process. It is striking to note that the focus is on the social dimension.

Table 4.10 Targets and target indicators of the 'Green Cabinet'

Generation Justice	Quality of Life	Social Bonds	International Responsibility
conservation of resources	economic prosperity	employment	development cooperation
climate protection	mobility	prospects for the family	opening markets
renewable energies	nutrition	equal rights	
land use	air quality	integration of foreign citizens	
biodiversity	health		
national indebtedness	crime		
economic precautions for the future			
innovation			
education			

Sustainable Development and Energy Supply

The adequate supply of useful energy forms, on the one hand, the basis for life in line with human dignity and an efficient society. On the other hand, the material flows associated with energy use dominate all the other material flows initiated by humans on the earth. The issue of future energy supply therefore occupies a prominent position in connection with the concept of Sustainable Development.

Present Status of the Discussion on Sustainable Development in the Energy Sector

Since the adoption of Agenda 21, the energy issue has been at the centre of the Rio process – either directly, when aspects of supply for humans are concerned, or indirectly, when dealing with the anthropogenic greenhouse effect. At the special session of the United Nations General Assembly in 1997 ('Rio plus 5'), the interdependence of Sustainable Development and the production, distribution and use of energy was emphasized once again. The General Assembly declared this topic to be a priority of the work of the United Nations Commission on Sustainable Development (CSD-9) in 2001. In preparation for the debate, the United Nations Development Programme (UNDP), the United Nations Department of Economic and Social Affairs (UNDESA), and the World Energy Council (WEC) organized a 'World Energy Assessment'. The results have been available since the year 2000. The associated report constitutes a comprehensive review of the social, economic and ecological aspects of energy supply and use, and of issues of supply assurance (UNDP, UNDESA, WEC: World Energy Assessment: Energy and the Challenge of Sustainability, 2000), but the report hardly addresses the structural change in the energy sector currently taking place at the company level.

In Germany, the Study Commission of the 11th German Bundestag 'Protection of the Earth' (1986-90) already devised proposals for a sustainable energy supply. It proceeded on the assumption that an energy supply permanently increasing by only 2 percent on an annual average would result in a 40 percent increase in CO_2 emissions directly attributable to the energy sector by the year 2005. The current emission level is expected to double by the year 2050.

The Study Commission on 'Protecting the Earth's Atmosphere' of the 12th German Bundestag (1990-94) essentially agreed with this estimate. Its basic concern was to improve approaches to significantly reducing CO_2 emissions. The commission formulated concrete reduction goals, which still represent a basis for the national climate protection programme today.

The Study Commission on 'Sustainable Energy Supply under the Conditions of Globalization and Deregulation' of the 14th German Bundestag (1999-02) orientated itself towards a target triangle composed of the corner points of safety, efficiency and environmental acceptability, which has long been the overall concept for energy economy and energy policy.

Dimensions of Sustainable Development in the Energy Sector

The top-down perspective of the concept of Sustainable Development implies that the dimensions of industry, society and ecology as the higher-level categories are also applicable to individual branches and different sectors such as the field of energy supply and energy use. However, with respect to their significance, they should be considered, interpreted, and substantiated in more detail. Consequently, overall concepts such as the above-mentioned target triangle of energy economy are to be subordinated to the concept of Sustainable Development. The results of this regulation and substantiation process will necessarily lead to repercussions on the higher levels. The overall concept of Sustainable Development is thus complemented by a bottom-up component.

At first sight, the demand for a more efficient energy supply can be directly related to the dimension of industrial development. The aspect of safe energy supply strongly corresponds with the social dimension, and environmental acceptability finds a counterpart in the ecological dimension.

The Connection between the Economic Dimension and Efficiency

Technical progress and economic growth are not only understood as the means for combating poverty in the developing countries, but are also regarded as the industrialized countries' necessary means to advance intergenerational justice. With the collapse of planned economies in the communistically governed states, a preliminary decision in favour of a competition-oriented system has been made with respect to a uniform global economic order. This basic order is also valid in the energy sector. However, there is resistance to specific forms of economic development. In particular, in connection with the globalization of economic processes, massive opposition is often expressed against an excessive pursuit of growth in the industrialized world, because, at least in the critics' opinion, the unfavourable situation in developing countries with thus further deteriorate.

Apart from the consequences resulting from deregulation and globalization, it is becoming apparent that an efficient use of scarce goods will gradually incorporate the 'environment' resource as well as the social resources. An example of environmental management is the future trade with certificates for emission rights in climate protection.

The promiscuous use of environmental goods often leads to environmental damage. This damage has so far not been allocated to the polluter, but is charged to third parties (as a rule, to the general public – and thus also to future generations). An internalization of external environmental costs is therefore considered to be 'a necessary condition in order to integrate the use of scarce environmental resources into the market process and subject it to the same management rules as the use of other scarce resources' (Study Commission: 'Sustainable Energy Supply under the Conditions of Globalization and Deregulation', 2001). However, relevant work is just beginning (European Commission: ExternE – Externalities of Energy, 1999).

The Connection between the Social Dimension and Safety

Two aspects can be distinguished in the social dimension of Sustainable Development: the first is related to the development policy discussion. Its main concern is to achieve intergenerational justice between countries at different levels of development. The primary aim is to make better development opportunities available to the poorer countries. Behind this is the thesis (or sometimes also the fear) that economic prosperity and social security could induce a more efficient and more conscientious handling of natural resources. In this connection, the persistent growth of world population and the possibility of extensive migration movements are named as threatening factors. In both cases, capacity building plays a central role. Its aim is to improve the prerequisites for Sustainable Development on site by improved education and further training, if possible, of all social strata in the developing countries. However, considerable financial resources are required. Existing instruments are not sufficient; and the responsible institutions do not appear to be adequately equipped for such tasks. The second aspect in relation to society deals with the social problems within an industrialized economy. The focus is on the prerequisites for social structural change, which is meanwhile considered necessary and is frequently oriented to ecological goals. Many of these changes are largely induced by economic processes. Problems arise especially whenever the pace of technical and economic change in a society exceeds what appears to be acceptable to its members.

The guarantee of a basic supply of energy services represents an important element in the debate on the social dimension of Sustainable Development. This applies to basic energy security – or basic energy supply – on the one hand, and on the other, to equal rights to use worldwide energy resources. Relative to the national context, questions must be raised here concerning the welfare state concept and the provisions for sickness and old age, the concept of individual responsibility and social security, as well as the achievement and solidarity principle.

In many countries on this earth, a deficient supply of energy induces a specific form of poverty (also called 'fuel poverty'). Consequently, the aim of sustainability-oriented strategies for the energy sector must be to remedy such injustices by means of suitable strategies as early and as far as possible. To this end, suitable complementary and compensation measures must be developed in a worldwide context. The deliberations under way prior to the Johannesburg conference show the extent of the problem addressed here.

Whereas from the developing countries' point of view, the question of basic supply is of major importance, aspects of supply security and operational safety come into play in the industrialized countries. Thus, the EU and Germany are greatly dependent on imported energy carriers for a foreseeable period. The discussion under way between Austria and the Czech Republic about the (safe) operation of the Temelin nuclear power plant points to the significance of operational safety and acceptance. In Germany, there are great differences at least concerning the question of whether the use of nuclear energy should be a component of Sustainable Development in the energy sector or not (see also Hake, 2001).

The Connection of the Ecological Dimension and Environmental Acceptability

If the normative postulate is applicable, according to which the coming generations must be granted options for satisfying their needs comparable to those available to the people currently populating the earth, it may be inferred that the earth (as well as the environment) must be put into a condition enabling people to satisfy their needs in the long run or that such a condition must be maintained. There are sometimes quite different views on how to structure the ecological heritage that will be left to future generations. Necessarily, between industrialized and developing countries there are considerable divergences in the ways in which these concepts are understood and evaluated. In general, it should be stated that the stress capacity or sustainability of the environment – i.e., the habitat of humans – is regarded as the essential ecological component for making effective the overall concept of Sustainable Development. These two 'values' constitute a suitable set of instruments for determining the regulation capacity of the total ecosystem within a foreseeable time frame. Stress capacity and sustainability of an ecosystem mean that this ecosystem tolerates a certain degree of (anthropogenic) stresses 'without its inherent or man-made structures and functions being changed' (Expert Council for Environmental Issues (SRU), 1994).

Emissions are generated during the extraction of coal, oil and natural gas, in the transport and distribution of natural gas, as well as during energy conversion in power plants and refineries, and the use of the fossil energy carriers such as coal, oil and gas in the final energy sectors (households, industry, traffic, trade and services as well as public institutions). These are primarily emissions of carbon dioxide and methane, nitrogen oxides, carbon monoxide and other volatile compounds. In total, the extraction, transport, conversion and use of fossil fuels globally contribute about 50 percent to the additional greenhouse effect (Study Commission: Preventive Measures to Protect the Earth's Atmosphere, 1991).

There is thus a correlation between an increased use of fossil energy carriers in the course of energy conversion and energy production, and increased emissions and environmental pollution. This correlation must be removed. Possible approaches include a further increase in efficiency (use of low-carbon and carbon-free energy carriers) and perhaps also large-scale technical separation and dumping of carbon dioxide (see also Kolb, 2001; Lackner, Ziock, 2001).

Not vi interested in Innovation issues !

Indicators and Guidelines for Sustainable Development in the Energy Sector

Indicators for Sustainable Development in the energy sector are components of a more complex higher-level system, where they play a central role, due to the significance of the energy sector. From the top-down perspective, a desegregation of indicators can be involved in a change to the next lower level. At the end of this multistage process there are frequently measurable indicators for which it is then possible to fix limit values in the sense of specifications as required.

The starting point is the assumption that the indicator sets derived for the different dimensions are universally applicable, taking into account, however, the different initial requirements of the respective countries. In what follows, we will describe the debate on indicators and guidelines for a Sustainable Development in the energy sector.

For the Federal Republic of Germany, the Study Commission on 'Preventive Measures to Protect the Earth's Atmosphere' of the 11th German Bundestag already pointed out in its report, 'Protection of the Earth', that an efficient handling of energy is indispensable for reducing environmental pollution.

The Study Commission 'Sustainable Energy Supply under the Conditions of Globalization and Deregulation' of the 14th German Bundestag established a catalogue of guidelines and instructions for a Sustainable Development of the energy economy. The commission's rules gave an equal ranking to ecological, economic and social interests. Moreover, the members voted for equal opportunities in accessing energy services, for conserving the natural environment and natural habitat, and for an increased use of renewable energies (Study Commission: Sustainable Development under the Conditions of Globalization and Deregulation, 2001).

Policy Targets ?

Furthermore, the study commission has designed its own set of indicators for the energy sector. It is based on the indicator systems of the OECD, the United Nations Commission on Sustainable Development, and on the national approaches of the Dutch, Norwegian, Swedish, British and Canadian governments. In addition to the OECD indicator set, the OECD considerations and Eurostat, the indicator concepts from the Netherlands, Canada, Norway, Sweden and United Kingdom have also been included in the analysis. The indicators for determining the ecological dimension should represent above all the urgency of the problem. This includes anthropogenic climate change, the emission of pollutants, the acidification of soils and waters, soil and landscape consumption, waste and refuse disposal, water pollution and biodiversity. In the field of economic indicators, the aim is to evaluate the different alternatives for the energy system by determining the costs. Another aim is to have better access to resources as well as to the resulting improvements in efficiency, thanks to better production and consumption patterns.

The indicators can be largely applied at the aggregated level of national economies. The study commission's indicator set highlights the relative contribution of individual technical and economic options (for instance, in energy supply) to sustainable development. At the same time, information about Sustainable Development is derived for the whole energy system and for its integration into the social and economic sector within the framework of a comparison of target/actual figures and via general development trends. A distinction is made here between key and auxiliary indicators, where key indicators denote those characteristics which aim at the respective outstanding trends and problems to be solved. In contrast, the auxiliary or additional indicators illustrate further aspects. Although these are not of such an outstanding significance as the key indicators, they may well expand the analysis. The study commission is aware that its indicator set – especially in the course of actual application – must be open to change. This applies, in particular, to the field of social indicators, where there is a real a lack of directionally reliable indicators.

Besides the study commission panel of the Parliament, the State Secretary Committee for Sustainable Development also published its first considerations on Sustainable Development for the Federal Government. In the field of energy supply and energy use, one of its major aims is to uncouple energy and resource consumption from economic growth. The energy-specific concept is entitled 'Using energy efficiently – protecting climate effectively'. This includes economic efficiency as a goal for producers and consumers, the conservation of the environment, climate and resources as well as the security of energy supply. Concrete fields of action are described by the Federal Government in the following areas:

- Increasing energy efficiency by reduced use of primary energies and a more efficient and economical energy use;
- More effective measures in the field of energy services concerning the proportion of industrialized to developing countries;
- Increased use of renewable energies – and as a countermove, energy-saving initiatives in regard to nuclear and fossil energy carriers.

The aim of sustainable climate protection and energy policy is manifested in the elements of the target triangle, which should be consistent with each other.

Furthermore, concrete measures and requirements are suggested in the draft of the State Secretary Committee for Sustainable Development (State Secretary Committee for Sustainable Development: Prospects for Germany, 2001). Among these goals favoured by the Federal Government are a 21 percent reduction of Kyoto gas emissions by 2012, a 25 percent reduction of CO_2 emissions by 2005 (compared to the base year 1990), a doubling of the share of renewable energies in total federal German energy consumption by 2010 (compared to the base year 2000), an expansion of combined heat and power generation, as well as an increase in energy efficiency in the private and industrial sectors, and a measurable increase in energy productivity. In the above areas, pilot projects have been initiated for an increased use of renewable energies and efficient energy via fuel cells.

The comparative analysis of different indicator systems on an international and national level give sustainable development a strong priority; and a sustainable power supply is obviously part of that. The allocation of the group of topics relating to 'Energy' is however not clear in this connection. This group of topics is allocated to the economic sector as well as to the environmental and social sectors. Remarkably, the majority of the industrialized countries normally view the power supply topic as belonging to the economic aspects of sustainable development. From a wider global standpoint, however, the power supply is related to the environmental dimension as well as to the distribution, i.e., political, dimension.

In the Sustainable Development context all aspects of energy supply and energy use are generally considered; but the focus is naturally on the establishment of the use of renewable energies.

Thus an intensified promotion of availability and use of renewable energies is the centre of attention, apart from the goal of the efficiency increase in conventional energy use. An important measured variable is thereby the range of the research funds made available both by national governments and by the industry for this sector. In this framework it is also necessary to measure the output of greenhouse gases. These are important indicators in reference to a successful conversion of the measures previously discussed.

It is of great importance that the selected single indicators of the different indicator models should be quantifiable. Thus the different indicator sets will be more readily comparable; and moreover, easily comparable default recommendations for action, as well as action measures, will be possible.

Conclusions

The currently debated concepts behind Sustainable Development have not yet been exhaustively discussed. It is striking to note that the approaches in most cases are not multidimensional. As a rule, one or another of the 'pillars' mentioned in the discourse is emphasized, excluding the others. This may be primarily explained by the complex structure of the overall concept of Sustainable Development. This can be confirmed especially for energy production and energy supply, where a one-sided approach is very often taken, exclusively privileging the economic or ecological dimensions.

In general, the debate on implementing the overall concept of Sustainable Development in the energy sector in Germany is characterized by two parallel developments. On the one hand, the debate is being conducted in a top-down fashion by government, other national institutions and international and supranational organizations (such as the OECD or the European Union). On the other hand, however, a bottom-up process involving societal initiatives can also be observed. This results in conflicts since the discussion about implementation of the overall concept moves on two different development lines. Whereas the national institutions and organizations concentrate on the socio-economic orientation of the sustainability concept in the energy sector, the groups involved in the bottom-up process generally vote for an increased environmental orientation of sustainability policy in the energy sector.

Consequently, at the present time, there is only limited social consensus regarding the overall concept of Sustainable Development in Germany. In particular, there is a conflict between the claim for an equal-ranking treatment of the economic, ecological and social dimension, on the one hand, and, on the other, emphasis on the primacy of ecology. These conflicts are at present also being discussed in the Study Commission 'Sustainable Energy Supply under the Boundary Conditions of Deregulation and Globalization' of the 14th German Bundestag.

In the course of the debate on the overall concept of Sustainable Development in the field of energy production and energy supply, the problem of risk assessment should be given more weight. In general, the idea of risk reduction and danger prevention by means of diversification (above all in the field of nuclear energy) with a view to ensuring supply seems to have been pushed into the background. However, there is the danger that one 'pillar' of previous energy supply is thus given up for no reason.

At the moment, capacity-building does not seem to be fully accounted for in the discussion on sustainability conducted in Germany at the federal level. Concerning the acceptance and applicability of the concept of Sustainable Development, society still seems to be in a learning phase. Increasing emphasis upon the responsibility of the population in this context (creation of 'energy awareness') thus appears appropriate, in addition to implementing the overall concept of Sustainable Development – as is currently under way in the Federal Republic of Germany in various policy areas and at various levels – as well as gradually anchoring the overall concept of Sustainable Development in the society, while grounding it within the political system.

At the same time, politicians, and above all political decision-makers in the federal and state governments, have the task, in addition to adapting the current policy fields and options, of opening up new fields and possibilities for action by speeding up efforts in the research and development sector. In this connection, an impartial discussion and consideration of the different opportunities and risks of possible alternative options in the energy production and energy conversion sectors appears necessary, with a view to developing and deriving alternative options.

Otherwise there is the threat that an excessively one-sided view will be represented, corresponding only to individual ideological demands and orientations. The debates and discussions about embedding the sustainability principle in the energy sector should be carried on in an open-ended fashion. All new developments and findings should be discussed and examined for the possibility of a successful integration into already existing and implemented policy approaches, and if possible, included.

In the midst of the debate about the integration of the concept of Sustainable Development into the energy sector, we may well ask whether a preference for the use of renewable energies in the future energy supply is really feasible. At the moment, it looks like this question has already been answered in the affirmative. However, this ignores the fact that such long-term opting-out scenarios from the 'conventional' energy supply are not yet available at all.

The findings of the 'Sustainable Energy Supply under the Conditions of Globalization and Deregulation' Study Commission of the 14th German Bundestag as well as the Federal Government's considerations with respect to a national sustainability strategy are just the beginnings of an effort to integrate the concept of Sustainable Development, and are more likely to raise new questions than answer already existing ones.

Sustainable Development implies ideas with roots in a new concept of society. The implementation of concepts corresponding to the local and global perspectives on the three dimensions of economy, society and environment, requires a common view of individual goals, values and measures.

Closely related to the debate on these issues is the issue of how the processes in question can be observed, measured and evaluated. Various indicator sets have been developed, and our review of the current ones demonstrates a considerable variety. Nor are the indicator sets perfectly free of bias; instead, each system usually reflects the political position of its author(s).

The indicator sets discussed so far have been mainly compiled by institutions from the industrialised world. Assuming that the less developed countries contribute to the process of establishing goals for Sustainable Development, the applicability of the indicator sets in these countries must be guaranteed; in other words, the countries have to improve their own capacities to measure the individual indicators. Standards are required to assure the same level of quality in order to reduce country-specific artefacts. For instance, the quality of drinking water should be measured by one set of indicators. The measurements should be based on the same procedure in all countries or regions.

It is also necessary to publish the values of the indicators to keep the public informed about the status of Sustainable Development. Depending on the progress of scientific understanding and the coefficients of the indicators, goals for Sustainable Development have to be revised or/and further action has to be taken. In many cases the energy sector can serve as a model for the implementation of strategies for Sustainable Development because the efficient, environmental friendly and safe supply of energy is fundamental for every economy.

References

Acker-Widmaier, G. (1999), *Intertemporale Gerechtigkeit und nachhaltiges.*Wirtschaften, Marburg.
BMVBW u. BMZ (2001), *Auf dem Weg zu einer nachhaltigen Siedlungsentwicklung. Nationalbericht der Bundesregierung zur 25. Sondersitzung der Generalversammlung der Vereinten Nationen ('Istanbul+5')* Berlin.
Bonse-Geuking, W. (2001), 'Nachhaltige Energiepolitik – Mehr Wettbewerb, weniger Staatseingriffe. Orientierungspunkte des Wirtschaftsrates der CDU e.V', *Energiewirtschaftliche Tagesfragen. Zeitschrift für Energiewirtschaft, Recht, Technik und Umwelt.* 12, December, pp. 769-771.
BUND und MISEREOR (ed.) (1996), *Zukunftsfähiges Deutschland. Ein Beitrag zu einer global nachhaltigen Entwicklung.* Basel, Boston, Berlin.
Bundesministerium für Umwelt, Naturschutz und Reaktortechnik (1992), *Agenda 21 – Konferenz der Vereinten Nationen für Umwelt und Entwicklung.* Bonn.
Bundesministerium für Umwelt, Naturschutz und Reaktorsicherheit (1997), *Auf dem Weg zur nachhaltigen Entwicklung in Deutschland. Bericht der Bundesregierung anlässlich der UN-Sondergeneralversammlung über Umwelt und Entwicklung im Juni 1997 in New York.* Bonn.
Bundesministerium für Umwelt (2000), *Naturschutz und Reaktorsicherheit: Erprobung der CSD-Nachhaltigkeitsindikatoren in Deutschland. Bericht der Bundesregierung.* Berlin.
Bundesministerium für Wirtschaft und Technologie (ed.) (2001), *Nachhaltige Energiepolitik für eine zukunftsfähige Energieversorgung. Energiebericht.* Berlin.
Bundesamt für Bauwesen und Raumordnung (ed.) (1999), *Nachhaltige Raum- und Siedlungsentwicklung – die regionale Perspektive.* vol. 7. Berlin.
Bundesamt für Bauwesen und Raumordnung (ed.) (2000), *Das Europäische Entwicklungskonzept und die Raumordnung in Deutschland.* vol. 3-4. Berlin.
Dollar, D. and Collier, P. (2002), *Globalization, Growth and Poverty: Building an inclusive world economy. A World Bank Policy Research Report.* Washington, D.C.
Eich, R. and Hake, J.-F. (2002), *Nachhaltige Entwicklung und Energie. In: Stein, Gotthard (Hrsg.), Umwelt und Technik im Gleichschritt. Systemanalyse und Technikfolgenforschung in Deutschland.* Jülich.
Enquete-Kommission 'Vorsorge zum Schutz der Erdatmosphäre' des Deutschen Bundestages (ed.) (1991), *Schutz der Erde. Eine Bestandsaufnahme mit Vorschlägen zu einer neuen Energiepolitik.* semi-volumes I and II. Karlsruhe, Bonn.
Enquete-Kommission 'Schutz des Menschen und der Umwelt' des Deutschen Bundestages (ed.) (1997), *Konzept Nachhaltigkeit – Fundamente für die Gesellschaft von morgen.* Bonn.
Enquete-Kommission 'Nachhaltige Energieversorgung unter den Bedingungen der Globalisierung und der Liberalisierung' des Deutschen Bundestages (ed.) (2001), *Nachhaltige Energieversorgung auf den liberalisierten und globalisierten Märkten: Bestandsaufnahme und Ansatzpunkte. Erster Bericht.* Berlin.
Enquete-Kommission 'Nachhaltige Energieversorgung unter den Bedingungen der Globalisierung und Liberalisierung' des Deutschen Bundestages (2001), *En route to sustainable energy use. Ninth Conference of the Commission on Sustainable Development* New York, 16-27. April 2001.
Enzensberger, N., Wietschel, M. and Rentz, O. (2001), 'Konkretisierung des Leitbilds einer nachhaltigen Entwicklung für den Energiesektor', *ZfE – Zeitschrift für Energiewirtschaft.* 25, pp. 125-137.
European Commission: ExternE (1999), *Externalities of Energy. Vol. 7. Methodology 1998 Update* (EUR 19083). Luxembourg.

European Commission: ExternE (1999), *Externalities of Energy. Vol. 8. Global Warming* (EUR 18836). Luxembourg.

European Commission: ExternE (1999), *Externalities of Energy. Vol. 9. Fuel Cycles for Emerging and End-Use Technologies, Transport and Waste* (EUR 18887). Luxembourg.

European Commission: ExternE (1999), *Externalities of Energy. Vol. 10. National Implementation* (EUR 18528). Luxembourg.

EU-Kommission (1993), *Wachstum, Wettbewerbsfähigkeit, Beschäftigung – Herausforderungen der Gegenwart und Wege ins 21. Jahrhundert. Weißbuch.* (KOM 93)700. Brussels, December.

Eurostat (1999), Towards Environmental Pressure Indicators for the EU. Ispra/Italy.

Forschungsverbund Sonnenenergie (ed.) (1999), *Nachhaltigkeit und Energie. Themen 98/99.* Köln.

Forschungszentrum Karlsruhe, Deutsches Luft- und Raumfahrtzentrum, Forschungszentrum Jülich, Gesellschaft für Mathematik und Datenverarbeitung, Umweltforschungszentrum Leipzig (2001), *HGF-Strategiefondsvorhaben 'Global zukunftsfähige Entwicklung – Perspektiven für Deutschland. Zwischenbericht'* Karlsruhe, Bonn.

Geiger, B. and Lindhorst, H. (2001), 'Energiewirtschaftliche Daten. Energieverbrauch in der Bundesrepublik Deutschland', *BWK. Das Energie-Fachmagazin.* vol. 1-2. pp. 40-44.

Görlach, B. (2000), *Evaluierung der Umweltintegration in der Europäischen Union im Rahmen der Studie 'Von Helsinki nach Göteborg'. Erarbeitet im Auftrag des österreichischen Bundesministeriums für Land- und Fortwirtschaft, Umwelt und Wasserwirtschaft.* Wien.

Hake, J.-F., Vögele, S., Kugeler, K., Pfaffenberger, W. and Wagner, H.-J. (eds) (1999), *Liberalisierung des Energiemarktes.* Forschungszentrum Jülich. Jülich.

Hake, J.-F., Vögele, S., Kugeler, K., Pfaffenberger, W. and Wagner, H.-J. (eds) (2000), *Zukunft unserer Energieversorgung.* Forschungszentrum Jülich. Jülich.

Hake, J.-F.(2001), 'Perspektiven für Kernenergie?' *GAIA.* Vol. 10. No. 1.

Hanekamp, G. and Steger, U. (eds) (2001), *Nachhaltige Entwicklung und Innovation im Energiebereich.* Graue Reihe – no. 28. Bad Neuenahr-Ahrweiler.

Hillebrand, B. et al. (2000), *Nachhaltige Entwicklung in Deutschland – Ausgewählte Probleme und Lösungsansätze.* Untersuchung des Rheinisch-Westfälischen Instituts für Wirtschaftsforschung. vol. 36. Essen.

IFOK Institut für Organisationskommunikation (ed.) (1997), *Bausteine für ein zukunftsfähiges Deutschland.* Wiesbaden.

Jochem, E. und H. Bradke (2001), 'Rationelle Energieverwendung: Von 5,5 kW auf 2,0 kW pro Kopf – die nachhaltige Industriegesellschaft', address in BMWi. Berlin, am 19. Juni.

Jörissen, J., Kopfmüller, J., Brandl, V. and Paetau, M. (1999), *Ein integratives Konzept nachhaltiger Entwicklung.* Karlsruhe, Bonn.

Kopfmüller, J., Brandl, V., Jörissen, J., Paetau, M., Banse, G., Coenen, R. and Grunwald, A. (2001), *Nachhaltige Entwicklung integrativ betrachtet. Konstitutive Elemente, Regeln, Indikatoren.* Berlin.

Maichel, G., Klemmer, P., Voß, A. and Grill, K.-D. (2000), *Leitlinien einer nachhaltigen Energiepolitik.* Sankt Augustin:.Konrad-Adenauer-Stiftung.

Mohr, H. (1995), *Qualitatives Wachstum.* Stuttgart.

Monstadt, J. (1997), 'Energiepolitik im Wandel zur Nachhaltigkeit?', *Berliner Beiträge zu Umwelt und Entwicklung,* Berlin: TU Berlin.

OECD (ed.) (1993), *Environment Monographs No.79. Indicators for the Integration of Environmental Concerns into Energy Policies.* Paris.

OECD (ed.) (1993), *Environment No. 83. OECD Core Set of Indicators for Environmental Performance Reviews.* Paris.

OECD (ed.) (1997), *Sustainable Development: OECD Policy Approaches for the 21st Century*. Paris.

OECD (ed.) (1998), *Towards Sustainable Development: Environmental Indicators*. Paris.

OECD (ed.) (2000), *Frameworks to Measure Sustainable Development*. Paris.

OECD (ed.) (2001), *Sustainable Development: Critical Issues*. Paris.

Rat für Nachhaltige Entwicklung (RNE) (2001), *Ziele zur Nachhaltigen Entwicklung in Deutschland – Schwerpunktthemen. Dialogpapier des Nachhaltigkeitsrates*. Berlin.

SDI Group (US Interagency Working Group on Sustainable Development Indicators) (1996), *Sustainable Development in the United States: An Experimental Set of Indicators*. Washington, D.C.

Redclift, M. (1987), *Sustainable Development. Exploring the Contradictions*. London.

Renn, O. and Kastenholz, H. (1996), 'Ein regionales Konzept nachhaltiger Entwicklung'. *GAIA*. Vol. 5. Nr. 2. pp. 86-102.

Rennings, K. and O. Hohmeyer (eds) (1997), *Nachhaltigkeit*. ZEW-Wirtschaftsanalysen. vol. 8. Baden-Baden.

Rennings, K. et al. (1997), *Nachhaltigkeit, Ordnungspolitik und freiwillige Selbstverpflichtung*. Heidelberg.

Schiffer, H.-W. (2001), 'Deutscher Energiemarkt 2000. Primärenergie – Treibhausgas-Emissionen – Mineralöl – Braunkohle – Steinkohle – Erdgas – Elektrizität – Energiepreise – Importrechnung'. *Energiewirtschaftliche Tagesfragen. Zeitschrift für Energiewirtschaft, Recht, Technik und Umwelt*. 3. pp. 106-120.

SRU (1994), *Umweltgutachten 1994. Für eine dauerhaft-umweltgerechte Entwicklung*. Mainz, Stuttgart.

Staatssekretärsausschuss für Nachhaltige Entwicklung (2001), *Perspektiven für Deutschland. Unsere Strategie für eine nachhaltige Entwicklung. Entwurf der nationalen Nachhaltigkeitsstrategie*. Berlin.

Umweltbundesamt (1997), *Nachhaltiges Deutschland. Wege zu einer dauerhaft-umweltgerechten Entwicklung*. Berlin.

UNDP (2000), *World Energy Assessment. Energy and the Challenge of Sustainability*. New York.

United Nations General Assembly (2000), *Report of the Secretary-General to the Preparatory for the High-level Intergouvernmental Event on Financing for Development*. New York, 18. December. (A/AC.257/12.)

Voss, G. (1997), *Das Leitbild der nachhaltigen Entwicklung – Darstellung und Kritik*. Deutscher Instituts-Verlag, Cologne.

Voß, A. (1999), 'Nachhaltige Entwicklung ohne Kernenergie?' *DatF* Winter 1999. 26. und 27. Januar 1999 in Bonn.

World Resources Institute (2001), *World Resources 2000/2001*. Washington, D.C.

Chapter 5

On the Creation and Distribution of Energy Rents

Bernard Beaudreau

[handwritten: Economics, Quebec eclectic 6-s'er]

In many, if not most, prehistoric societies, the sun was regarded as a god, the source of all life on earth. Sun gods included Apollo (Greek and Roman), Cautha (Etruscan), Shamesh (Sumerian), Dhatar (Hindu) and Knich Kakmo (Mayan), to name a few. Interestingly, the word deity has its etymological roots in the Indo-European dyeu, or god of light (energy). In the mid 1700s, the French Physiocrats argued that agriculture was the source of all material wealth, and more importantly, the only source of economic surplus. More specifically, agriculture, unlike all other sectors of the economy, was the only one capable of producing a surplus.

Today, these ideas are largely forgotten and/or ignored. Developments in the physical sciences, however, have corroborated ancient notions of the role of the sun and energy in the organization of life on our planet, and, indeed, elsewhere in the universe. Physicists, for example, refer to the four fundamental fields (forces) of the universe (strong force, weak force, electro-magnetic force, and gravitational force). All material processes, including photosynthesis, are powered by one (or more) of these forces. Given the near absence of non-muscle-based energy forms in the 18th century, the basic underlying notion of Physiocratic thought can be regarded as being consistent with modern physics. For example, artisanal production, being human-muscle based, is ultimately fuelled by the sun via photosynthesis. Put differently, solar radiation is converted, via photosynthesis, into sugars and starches, fuelling muscular force, and is, ultimately, the force behind the creation of material wealth.

Continuing along these lines (reductionism), it could be argued that in Physiocratic thought, energy is the source of all rents (surplus). One could define these rents as the difference between total potential energy as represented by the available food in any given year, and the energy required to sustain life in that year. Any surplus could then be used to generate more wealth, say via better nutrition, and the resulting increased physical (muscular) effort.

The theory of energy rents was developed in the late 1990s (Beaudreau, 1998). Using the energy-organization model of production, it was shown that energy consumption in production processes generated considerable energy rents, defined as the difference between the marginal revenue product of energy and its price (cost). As labor is no longer a source of energy/force in modern production processes, and capital is, by its very nature, not physically productive (classical mechanics), both were redefined as organization-related factor inputs. That is, factor inputs that, together, define the entropic framework. Other organization-related inputs include information and management. In short, the energy-organization approach to modelling production processes focuses on two factor input groupings, namely broadly defined energy and broadly defined organization.

The problem of distribution is relatively simple, and consists merely in apportioning energy rents. Energy, in a broadly defined sense, is the only physically productive factor input; therefore, for a given material process, it can be thought of as the sole source, and therefore also as the cause, of wealth. Such a view, however, ignores the role of organization. After all, energy without organization would be wasted. Organization, broadly defined, is thus a productive factor input; however, it is not a physically productive one. Accordingly, it is entitled to a share of the spoils.

As defined, energy rents are the source of material wealth, the latter being defined in turn as the aggregate value of all material transformations. By netting out the cost of energy (extraction, transportation, distribution, etc.), one obtains aggregate gross domestic product (gross national product). It follows that to increase income (material wealth), a society must generate additional energy rents, which, as pointed out above, implies increased levels of energy and organization (as factor inputs).

The rate of growth of energy rents is, as such, an increasing function of the rate of growth of energy consumption, and the rate of growth of organization, broadly defined. As pointed out in Beaudreau (1995, 1998, 1999), the productivity and growth slowdowns of the 1980s and 1990s can be directly attributed to the marked slowdown in the rate of growth of energy consumption among industrialized nations. It is worth noting that in February 1973, oil prices quadrupled, as the result of the OPEC oil embargo. Since production processes are by definition material processes, it stands to reason that higher energy costs (and the prospect of even-higher energy costs) would reduce the rate of growth of energy use and, as such, the rate of growth of work (output). Concomitant with the slowdown were a number of well-known developments, including, the slowdown in the rate of growth of per-capita output and, as such, real wages, the information and communications technology (ICT) revolution, globalization, an increasingly unequal distribution of income, and overall economic stagnation. The economic growth and stability of the post-WWII period gave way to two decades of weak growth, virtually no per-capita output growth and, not surprisingly, rising social and political tensions.

In this chapter, these developments are examined through the prism of the energy-organization approach to modelling production processes, and the resulting energy-rents approach to understanding income distribution. Thus far, there have been relatively few attempts at providing an integrated account of this misunderstood, yet extremely important period of post-WWII world history. For example, it is shown that globalization and the ICT revolution came on the heels of a marked decrease in the rate of growth of energy rents. At this point in time, income distribution, until then a non-zero sum game, became a zero-sum game. Higher wages (real wages) resulted in lower profits, which in turn prompted firms to accelerate ICT-related investment and invest offshore. The latter two strategies can, as such, be seen as responses to the new reality that was zero energy rent growth.

The chapter is organized as follows. In the next section, the theory of energy rents developed in Beaudreau (1998) is used to examine economic growth and income distribution in the post-WWII period. Per-capita income growth is shown to be highly co-linear with per-capita energy consumption, as measured by per-capita electric power consumption. Particular emphasis is placed on the distribution of energy rents, both at the individual firm level and at the aggregate level. It is argued that taxation constitutes a means whereby energy rents are distributed more equally among individuals. Armed with these findings, we then turn and examine the fallout of the energy crisis, paying particular attention to firm behaviour. Globalization, and automation are two of the strategies examined. It is argued that globalization and automation resulted from two related phenomena. First, energy rents stopped growing, the result of the energy crisis. Second, rising real wages in the presence of the former (i.e. zero energy rent growth) contributed to lowering real profits. Together, these two phenomena contributed to globalization and automation. The last section examines the physical (read: scientific) underpinnings of the energy-rent approach to material wealth, and the implications for economic growth and development in general.

Energy Rents: The Evidence

The concept of energy rent was an outgrowth of my theoretical work in the 1990s in the field of production theory (Beaudreau 1995, 1998, 1999). Responding to the paucity of models of production activity-incorporating energy in a meaningful and realistic way, I developed the energy-organization framework. Drawing from the related fields of biology and engineering, material processes were formalized in terms of broadly defined energy and organization. (This view of production is analogous to that found in Soddy [1924] and in Scott [1933].) Work, measured in terms of value added, is an increasing function of force (energy). Mediating the relationship between these two variables is the concept of second-law efficiency, or the extent to which a given amount of force can perform work. This is formalized in terms of Equation 5.1 where $W(t)$ = work in period t; and $E(t)$ = energy (i.e. force) in period t, η = second-law efficiency, $S(t)$ = supervision in time t, and $T(t)$ = tools (machinery and equipment) at time t, respectively.

$$W(t) = \eta[S(t); T(t)]E(t) \tag{1}$$

Second-law efficiency (i.e. η) is assumed to require an increase in tools and supervision. Capital and supervision are not productive in the traditional (read: physical) sense; rather, they contribute to work via organization, approximated here by second-law efficiency. The better capital is, the better supervision is, the higher η is, and hence, the higher quality the work is. Supervisory activity is traditionally organized hierarchically, with workers (lower-level supervisors) at the bottom, line supervisors above, and senior managers (e.g. CEO's, CFO's and the Board of Directors) at the top.

While not a source of energy, the supervisory input is nonetheless a *sine quo non* of virtually all production processes. Without supervisors (lower and upper level), work becomes probabilistic (including the null set). Machinery and equipment could break down, resulting in a loss of energy and output. Lack of – or breakdown in – organization is, theoretically equivalent to a value for η which tends toward zero.

Energy Rents

As pointed out earlier, energy rents are, by definition, the difference between the marginal revenue product of energy and its cost. As organizational factor inputs are not physically productive, it stands to reason that energy rents can be formally defined as $P \eta E(t) - P_{E(t)}E(t)$, where P and $P_{E(t)}$ represent the price of value added (work) and the price of energy, respectively. This corresponds to the surplus value (analogous to value added) physically produced by the energy input. This is analogous to the Marxist notion of labor surplus value, the difference being the source of value, which in this case is energy.

The notion of energy rent, I argue, underlies the Physiocratic belief that agriculture was the only sector in the economy that could generate a surplus. In a world of animate, muscular energy (that of pre-industrial France), carbohydrates and proteins define the overall energy constraint. The human body, one must keep in mind, simply transforms food energy into muscular force, which, in turn, is used to transform materials (artisans). Therefore solar radiation is the ultimate energy source, and, more importantly, the only possible source of surplus economic value radiation. This view also provides the wherewithal to rationalize the practice of worshiping sun gods. Quite simply, solar radiation is the ultimate source of all wealth.

Energy Deepening and Energy Rents: The Historical Record

British Prime Minister David Lloyd George, writing in the *Coal and Power Report*, referred to the relationship between energy deepening and energy rents in the following way:

> Those people are best paid and most prosperous that make most use of the resources of science ... the average level of earnings must depend on production and production increases as the use of power per head of population increases (David Lloyd George, 1924).

Accordingly, per head production is an increasing function of the resources of science, which, in this case, refers to steam and/or electric power. The *Coal and Power Report* was released in 1924 and called for a radical reorganization of the production and distribution of electric power in Great Britain. The US experience with electrification weighed heavily in the minds of Lloyd George and others. Great Britain would have to shift from steam power to electric power.

In this section, I examine the historical record of energy consumption and energy rents. Given the paucity of historical data prior to the Great Depression (1930s), the focus will be on the post-WWII period. Making use of the data on post-WWII Germany, Japan and the United States, and employing the LINUX estimation technique, Kummel, Eichorn and Lindenberger (1998), examined the relationship between energy use and manufacturing output. In similar work, Beaudreau (1995, 1998), using the method of constrained least squares, estimated a similar relationship, again using data from US, German and Japanese manufacturing. All three studies reported output elasticities for energy (electric power) in the range of 0.50 to 0.60, providing support for the energy-organization approach to modelling material processes. For example, Table 5.1, taken from Beaudreau (1998), shows energy output elasticities ranging from 0.4929 in the case of the United States to 0.7474 in the case of Germany.

Table 5.1 Output elasticities: US, German and Japanese manufacturing

Independent Variables	US	Germany	Japan
EP	0.492943 (26.551)	0.747482 (3.135)	0.605599 (3.017)
L	0.413503 (18.231)	0.121134 (2.332)	0.197653 (1.847)
K	0.093553 (2.768)	0.131383 (0.543)	0.196748 (1.608)
Constant	0.080902 (9.956)	0.046106 (1.426)	-0.019274 (0.271)
R2	0.99152	0.95821	0.98314
F	2279.2802	229.2853	612.1780

These estimates, in combination with data on actual factor shares, provide empirical support for the presence of energy rents. Reported factor shares for energy in manufacturing range from 0.04 to 0.06. That is, the share of energy in overall manufacturing costs ranges from 4 percent to 6 percent. The difference between output elasticities and factor shares, we argue, constitutes evidence of energy rents. Put differently, the contribution of energy to overall output exceeds, in a non-negligible way, the overall cost of energy.

Income Distribution as a Cooperative Bargaining Game

As pointed out in Beaudreau (1998), the laws of classical mechanics preclude the use of physical productivity as the basis for income distribution. More to the point, as neither capital nor labor is physically productive in modern production processes, existing factor shares, whether at the firm, industry or national level, cannot be rationalized on the basis of physical productivity. As energy, in its broadly defined sense, constitutes the only physically productive factor input, a purely physical productivity standard would have all of the output going to the owners of energy, and none to the owners of broadly-defined organization (labor and capital).

    To get around this problem, Beaudreau (1998) developed a model of income distribution based on energy rents and cooperative bargaining theory. Specifically, the problem of income distribution can be understood as a cooperative bargaining game involving the owners of energy and organization over the resulting energy rents, per se. Income distribution, as such, is defined as the solution to the underlying game. Among the factors affecting the solution is each player's bargaining power, as well as each player's outside option.

Bargaining Without Outside Options

Let us begin by defining the bargaining problem. The owners of energy and organization (e.g. the owners of energy, tools, the supervisory inputs, and lastly, the owners of the production processes themselves) bargain over $W(t) = \eta[S(t), T(t)]E(t)$, the output, in this case, manufacturing value added. In this case, $W(t)$ is expressed as units of a good/service. Later, it will be converted to a money-value equivalent. Define s_E, s_K, s_{S_l}, and s_{S_u}, where $[0 \leq s_i \leq 1, \Sigma_{i=E;K;S_l;S_u} s_i =1]$, as the electric power, capital, lower-level and upper-level factor share, respectively. Also, assume that α_i where $i= E, K, S_l$ and S_u, defines factor i's bargaining power $[0 \leq \alpha_i \leq 1, \Sigma_{i=E;K;S_l;S_u} \alpha_i=1]$. Lastly, assume that factor i's utility is an increasing linear function of income. More specifically, $U_i = U_i[s_i W(t)] \ \forall \ i = E, K, S_l, S_u$.

    This provides a convenient framework in which to study income distribution. In the absence of outside options, the simple bargaining problem is given by Equation 5.1, where the s_i $i = E;K;$ $S_l;$ S_u are chosen to maximize Equation 5.2, which, by definition, consists of the inner product of each factor owner's utility function.

$$\max_{s_i} S = \prod_{i=E,K,S_l,S_u} [s_i W(t)]^{\alpha_i} \tag{2}$$

    If we let $\alpha_i = 1/4$, then the solution to this problem is given by $s_i=1/4$ for all $i=E;K;$ $S_l;$ S_u. In this case, energy rents are, by definition, $75W(t)$, or three-quarters of overall value added. Thus, in a world devoid of outside options and identical preferences (i.e. across factor owners), income distribution will be determined by relative bargaining power. For example, the greater the lower-level supervisors' bargaining power is, the greater there share of the pie is, so to speak.

Bargaining with Outside Options

The presence of outside options alters considerably the bargaining problem. For example, let us suppose that the owners of electric power can sell each kilowatt-hour at a price of 7 cents. It stands to reason that, at the very least, the owners' share of manufacturing output must be equal to or greater then the corresponding market value of the power. Define ξ_i such that $\xi_i > 0$ is factor i's outside option. The bargaining problem becomes:

$$\max_{s_i} S = \prod_{i=E,K,S_l,S_u} [s_i W(t) - \xi_i]^{\alpha_i} \tag{3}$$

subject to:

$$s_i W(t) - \xi_i \geq 0 \forall i = EP, K, S_l, and S_u \tag{4}$$

In this case, a bargain (income distribution) will be struck if and only if the various factor inputs receive, at the very least, their outside options; otherwise, negotiations will break down, resulting in no output (value added, work). It therefore follows that Equation 5.3. must hold for all $i = EP$, K, S_l, and S_u.

The Determinants of Outside Options

What determines each factor input's bargaining power? What determines each factor input's outside option? Outside options refer to the value of alternatives, in this case, alternatives to entering into an agreement to produce value added. For example, the owners of electric power could conceivably consume their kilowatt-hours instead of producing added value. The owners of capital (tools) could conceivably consume their capital. Lastly, the owners of upper and lower-level supervisory skills could conceivably devote all of their time to leisure activities. In a world in which there is more than one firm, the owners of these factor inputs could, theoretically, bargain with a number of firms.

The Determinants of Bargaining Power

For the set of bargaining problems where the number of potential solutions is greater than one (i.e. the perfectly competitive bargaining solution, defined by a strict equality for Equation 5.3), bargaining power plays a crucial role in income distribution. (In the perfectly competitive case, energy rents are zero.) For example, the more bargaining power the owners of supervisory inputs have over the owners of electric power, the greater their share of the final product will be.

This raises the question of bargaining power per se. What determines the distribution of bargaining power among the owners of electric power, capital, and supervisory inputs? Unfortunately, while the bargaining model describes the process of income distribution in the presence of rents (solution to the game), it comes up short in so far as the determinants of bargaining power are concerned. More to the point, bargaining power is parametric to the model.

Energy and Culture

Table 5.2   Value added earnings per production worker, US manufacturing 1958-93

Year	VAPW*	SALPW*	Ratio
1958	12,308.04	4,246.62	0.34502
1959	13,152.48	4,388.02	0.33362
1960	13,206.89	4,404.82	0.33352
1961	13,547.86	4,445.06	0.32810
1962	14,177.45	4,609.10	0.32515
1963	14,874.40	4,735.37	0.31835
1964	15.510.90	4,878.83	0.31454
1965	15,873.49	4,921.01	0.31001
1966	16,150.46	4,969.17	0.30768
1967	16,189.97	4,959.56	0.30633
1968	16,833.30	5,094.24	0.30262
1969	16,759.92	5,073.52	0.30271
1970	16,665.35	5,008.73	0.30054
1971	17,582.53	5,135.34	0.29207
1972	18,253.98	5,353.77	0.29329
1973	18,720.17	5,373.61	0.28704
1974	19,275.58	5,226.61	0.27115
1975	19,224.78	5,155.51	0.26817
1976	20,242.33	5,317.70	0.26270
1977	20,759.87	5,440.70	0.26207
1978	20,867.14	5,458.59	0.26158
1979	20,864.07	5,249.91	0.25162
1980	18,793.47	4,968.26	0.26436
1981	20,124.74	4,948.33	0.24588
1982	20,433.51	4,914.09	0.24049
1983	21,514.38	5,018.42	0.23325
1984	22,322.87	5,097.87	0.22836
1985	22,627.40	5,169.91	0.22848
1986	23,573.36	5,230.92	0.22190
1988	25,330.14	5,165.33	0.20392
1989	25,248.55	5,070.17	0.20081
1990	24,978.22	4,995.13	0.19998
1991	25,096.23	4,959.87	0.19763
1992	26,098.23	5,029.40	0.01927
1993	26,230.17	4,999.69	0.19060

* Constant 1958 dollars

Income Distribution in US Manufacturing 1958-93

In this section, the bargaining approach to income distribution is used to analyse income distribution in US manufacturing from 1929-84, a period characterized by increasing energy deepening (until the mid-1970s). Energy deepening, defined as an increase in the energy/labor and energy/capital ratio, increased the overall level of energy rents, over which the owners of labor and capital (organizational inputs) bargained. The result was higher real wages and real profits, as organizational inputs (labor and capital) appropriated the resulting energy rents. This process, however, came to an abrupt end in the late 1970s. The energy crises had put an end to over six decades of energy deepening (for more on the energy crises, see Beaudreau [1995, 1998]).

In keeping with the predictions of the model, it can be argued that actual factor income shares mirror relative bargaining power. Referring to Table 5.2 which reports the share of labor (production workers), expressed as the ratio of two ratios, namely the ratio of value added to production workers in US manufacturing as reported by the US Bureau of Census, Survey of Manufacturing, and the ratio of production workers' overall earnings (wages) to production workers. We see that from a high of 34 percent in 1958, this ratio has decreased monotonically, experiencing its greatest decrease from 1973 onward, when it fell from 28 percent to 19 percent. One can argue that this decrease mirrors labor's diminished and diminishing bargaining power.

Energy throughout this period received roughly 6 percent of value added. This stands in stark contrast with energy's actual physical productivity. Energy in the form of electric power, the only physically productive factor input, received a mere 4 percent of total income. On the other hand, labor which, over the course of the past two centuries, has been reduced to a supervisory input (i.e. lower-level supervisors) in manufacturing production processes received 65 percent.

Income Taxation

That income distribution cannot be rationalized in terms of physical productivity has a number of implications in so far as public policy is concerned. For example, there is the question of equity, more specifically, equity as it relates to energy rents. How should energy rents be distributed? Who owns broadly defined energy and, consequently, who can legitimately lay claim to the resulting energy rents? These issues were raised by Buckminster Fuller in his work on cosmic energy accounting. As he pointed out:

> Humanity's cosmic-energy income account consists entirely of our gravity-and star (99 percent Sun)-distributed cosmic dividends of water power, tidal power, wave power, wind power, vegetation-produced alcohols, methane gas, vulcanism, and so on. Humanity's present rate of total energy consumption amounts to only one four-millionth of 1 percent of the rate of its energy income (Fuller, 1981, p. ii).

Is bargaining power, as defined above, a legitimate and equitable criterion? That is, can income distribution be legitimized in terms of bargaining power?

Clearly, these are major issues, issues that are beyond the scope of the present paper. It is worth noting, however, that the concepts of energy rents and bargaining presented here provide an additional rationale for progressive taxation, namely as a means of sharing, in an equitable fashion, the energy rents that are generated in any given society. As the bulk of wages and profits of energy rents are obtained via the bargaining process (not earned in the physical sense), the question of equity arises. As no one individual or group of individuals (workers, capitalists) can legitimately lay claim to energy (ownership), it could be argued that the government should oversee the distribution of energy rents. That is, as the energy rents belong to all, it stands to reason that the government should oversee the distribution of energy rents. Among the available policy instruments are (*i*) policies affecting relative bargaining power, and (*ii*) policies affecting the actual distribution of income. The former are indirect in the sense that they attempt to affect income distribution at the source (i.e. at the firm level). The latter are direct policies as they deal directly with income (earned).

Examples of policies affecting bargaining power include labor-related legislation such as the National Labor Relations Act (1935) that created the National Labor Relations Board in the United States. In short, this piece of legislation provided labor with the wherewithal (i.e. collective bargaining) to negotiate higher real wages (greater share of energy rents). Similar laws were enacted in nearly all industrialized nations in the 1930s, a period characterized by massive energy deepening (electrification), and similar labor legislation was also enacted in less-developed countries such as Mexico.

Examples of policies affecting the actual distribution of income include income taxation and redistribution, energy taxation, and indirect taxation. Interestingly, income taxation on a large scale, and redistribution policies, date back to the same period, namely, the 1930s. With increasing energy rents, governments in most industrialized nations innovated with progressive taxation and redistribution. In doing so, those appropriating above-average energy rents would be required to give up a fraction that would then be given to those with below-average energy rents. When the owners of capital avoided taxation by choosing to reinvest earnings at the firm level, governments resorted to a tax on profits. The Undistributed Profits Tax of 1936 in the United States is an example.

Energy Rent Stasis: The Fallout

With the energy crises of the 1970s came the end of energy deepening, energy rent growth, and higher real wages and profits. High energy costs, in combination with the specter of even higher energy costs, put an end to energy deepening. Whereas prior to the 1970s, process engineers had focused on higher throughput rates (energy deepening), from the 1970s on, they would focus on energy efficiency. The resulting fallout is commonly referred to as the productivity slowdown. Throughout the 1980s and 1990s energy consumption grew at historically low rates, resulting in low rates of output and productivity growth.

Table 5.3 reports growth rates for manufacturing value added (*Q*), electric power (*E*), labor (*L*) and capital (*K*) for three time intervals: 1950-84, 1950-73, and 1974-84 in the US, Germany and Japan. Energy consumption growth virtually plummeted in all three countries, going from 6.226 in the pre-energy crisis period to 0.330 percent in the post-energy crisis period in the US, from 5.883 percent in Germany to 1.366 percent, and from 11.32 percent to 0.965 percent in Japan.

The fallout was cataclysmic, to say the least. The West had grown accustomed to high growth, high energy rents, high wages, and high profits. Year in and year out, workers could expect their real income to rise by 2-3 percent, as could the owners of capital. From the mid-1970s on, overall output and labor productivity growth came to a halt.

Table 5.3   Output and input growth rates: US, German, and Japanese manufacturing

	U.S.		
	1950–1984	1950–1973	1974–1984
Q	2.995	4.217	0.330
EP	4.455	6.226	0.591
L	0.784	1.375	-0.503
K	3.564	3.651	3.378
	Germany		
	1962–1988	1962–1973	1974–1988
Q	3.054	4.954	2.700
EP	2.894	5.883	1.366
L	0.834	1.703	0.344
K	4.082	5.199	3.226
	Japan		
	1965–1988	1965–1973	1974–1988
Q	3.826	8.844	3.099
EP	3.559	11.320	0.965
L	-0.082	2.297	-0.367
K	7.520	13.536	5.182

$*\alpha_1 \frac{\dot{ep}(t)}{ep(t)} + \alpha_2 \frac{\dot{k}(t)}{k(t)} + \alpha_3 \frac{\dot{l}(t)}{l(t)}$, where $\hat{\alpha}_i$ are the estimated coefficients.
Source: Beaudreau (1995a,b,c).

While actual growth may have come to a halt, growth expectations remained buoyant. Workers continued to expect wage increases in the order of 2-3 percent. Rising energy prices, on the other hand, contributed to reducing energy rents, with predictable results. The solution to the bargaining game referred to above could no longer be supported. The consequences were severe, and included increased labor unrest, as measured by increased strike activity, lower stock prices, reflecting what appeared to be shrinking growth opportunities, and, in the longer run, a number of responses, ranging from increased automation to off-shore relocation of labor-intensive and automation-resistant manufacturing activity. Higher energy costs, and the specter of even higher energy costs prompted firms to take steps to restore the bottom line, and, hopefully, restore earnings growth.

Automation and Energy Rents

Automation can be defined as the substitution of animate (i.e. human being-based) lower-level supervisors by inanimate lower-level supervision. In short, highly sophisticated control devices, mostly based on the microchip, take the place of workers (supervisors). Automation was not new to the 1980s and 1990s, if one considers the historic origins of modern automation with the Jacquard loom in 18th-century Great Britain. The energy crisis, and the resulting violation of the cooperative bargaining solution to the problem of income distribution, I maintain, contributed to increasing the rate of diffusion of this new technology. Automated

uses all his own terms!

manufacturing systems, better known as intelligent manufacturing, and CAMI (computer-assisted manufacturing industry) soon became the norm.

From a financial point of view, automation was a win-win proposal. In the limiting case in which a firm could reduce its labor force entirely, capital and management could appropriate labor's share of the firm's energy rents. The costs of automation are negligible; specifically, the variable costs are negligible, consisting of the cost of powering the control devices (computers).

It is important to point out that automation did not increase overall output, for obvious reasons. As pointed out above, organization is not a physically productive factor input. Automation did, however, contribute to increasing conventionally defined labor productivity. The latter is defined as the ratio of overall output to labor input. Falling labor input would, in the absence of an increase in overall output, still contribute to increasing labor productivity.

Unfortunately, not all production processes can be automated. Take, for example, the textile and garment industry where production processes remain labor intensive. Computer and robot technology have not yet succeeded in eliminating garment workers (lower-level supervisors). They are sources of both energy and organization, owing to the nature of the underlying production processes. Piecing together an article of clothing requires both inanimate (i.e. non-muscular) energy in the form of electric power, and animate energy in the form of arms, hands and fingers to manipulate the cloth.

Globalization and Energy Rents

Globalization as seed for energy rents!

In this section, I maintain that globalization can best be understood in terms of energy rents, more specifically, in terms of energy rent growth. As pointed out, the 1980s witnessed the end of energy deepening, and energy rent growth. Higher energy prices, in combination with workers' wage expectations, rendered the solution to the cooperative bargaining problem indicated above untenable. This prompted a number of responses, including automation, and where not feasible, offshore investments. Furthermore, attempts were made by Western industrialized nations to find new, cheaper supplies of energy (read: fossil fuels). If successful, the latter would provide the wherewithal to resume energy deepening.

Offshore investment (up-stream vertical foreign direct investment) provided an ideal alternative to automation for firms in automation-resistant sectors. Firms in these sectors sought out locations offering cheap (and more importantly, readily-available energy), and cheap labor. When combined with foreign tools and upper-level supervision, the result was higher profits.

Globalization, as such, can be seen as an attempt on the part of firms – more specifically, the owners of capital and upper-level supervisors – to increase their share of energy rents. In the new locations, production processes continued to be powered by inanimate energy in the form of electric power, the only difference being that foreign workers received a smaller share of the energy rents.

A good example is the garment and textile sector, where offshore relocation was accomplished on a massive scale in the 1980s and 1990s, with the result that world production is today concentrated in low-wage, energy-abundant countries.

Another example is the electronics industry, or more specifically, the electronics assembly industry, which is situated almost exclusively in low-wage, energy-abundant countries. In both cases, Western firms simply relocated plant and equipment to low-wage countries. The technology, as such, was identical. What did differ, however, was the factor shares (i.e. the $s_i \forall i=EP; K; S_l; S_u$).

Globalization and Energy Rent Growth

Offshore relocation, as pointed out, was a profit-augmenting strategy. The problem, however, is the finite nature of such gains. Once all automation-resistant production processes are relocated to low-wage countries, profit opportunities will cease, for obvious reasons. This differs markedly from profit growth in the pre-energy crisis era. In the latter period, profit growth resulted from energy deepening, in the form of higher energy to labor and energy to capital ratios. Put differently, the energy rent pie was constantly growing, providing for bigger and bigger slices.

Seen from a different light, the energy crises had the effect of transforming what had, until then, been a non-zero-sum game bargaining problem, into a zero-sum game bargaining problem. This change applied to income distribution as well. Capital's gains would from that point on be achieved at the expense of labor.

The problem was with energy, specifically, with energy deepening. For the West to return to consistently high growth rates, it would have to find new, abundant sources of energy. Increased capital, labor, information, etc. would not suffice, as these are organization factor inputs (read: not physically productive). As pointed out in Beaudreau (1999), organization factor inputs can only affect the rate of growth of output via second-law efficiency – that is, indirectly. Better tools and better supervision (i.e. the use of control devices) can raise second-law efficiency. A good example of higher η is the set of alloys known as superconductors that conduct electricity without resistance, thus increasing transmission efficiency. However, the potential gains are limited, as the very nature of second-law efficiency is noted for being highly stable.

Growth Perspectives

The energy-organization framework outlined above can be used to examine the issue of future growth perspectives. What is the likelihood that Western industrialized economies will grow at pre-energy crisis levels? Are pre-energy crisis growth rates possible? Unlike competing approaches, where the emphasis is on factors such as research and development, the energy-organization approach puts the emphasis on energy growth, more specifically, on the rate of growth of energy consumption. The reason is relatively simple, namely that while organization-related inputs are necessary conditions for growth, their contribution is relatively stable (second-law efficiency).

The ICT Revolution

The development of the microchip in the 1960s laid the foundation for an information revolution based on cheaper information storage and retrieval. Information could be stored and retrieved at a near-zero cost. To many, the microchip was as important as the steam engine and the electric motor – if not more important. Like the first and second industrial revolution both of which resulted in manifold increases in output, these writers predicted a manifold increase in productivity and output.

Thus far, these predictions have not been borne out. Productivity and output growth rates have not increased as expected. The technology – i.e., information and communication technology – bubble has burst. Economists have responded by examining the underlying causes, paying particular attention to dissemination lags or delays. For example, Paul David (1990) has argued that the case with the dynamo (electric motor), whose affect on productivity and output was not instantaneous but rather spread over an extended period of time, also applies to the ICT revolution, which would be years in the making. More recently, Edmund Phelps (2004) pointed to the similarity between the stock market boom and crash in the 1920s, and recent stock market boom and crash.

The limit of this literature, and indeed of the whole growth literature, is the lack of a theory of production that explicitly incorporates information in a meaningful way. Rather than simply including information alongside capital and labor in standard production functions, an attempt should be made to examine the exact role of information in material processes. Is it physically productive, and, if so, how?

The energy-organization framework presented earlier addresses these and other questions regarding information. Drawing from classical mechanics and thermodynamics, it sees information as an organization-related factor input. As such, it, like capital and labor, is organizationally productive, but not physically productive. This follows from the fact that it is not a source of energy. Better information will, as such, serve to increase second-law efficiency; however, more of the same information will be of little to no incremental value.

This has a number of implications for the cause of the 'third industrial revolution'. To begin with, because information is not a physically productive factor input, the ICT revolution cannot and, more importantly will not, caeteris paribus, contribute to an increase in the growth rate of productivity and output. Second, information and energy are not comparable; hence, comparisons between the first, second, and third industrial revolutions are of no apparent consequence. Lastly, the recent stock market crash (technology sector) is not an anomaly, but, instead, perfectly consistent with classic mechanics and thermodynamics. Unfortunately, information has not been productive, is not productive, and will never be productive.

The Spectre of Cheap Energy

Output growth without energy growth violates basic classic mechanics, thermodynamics and virtually all of physics. Production processes are material processes, which, like all material processes, are energy based. It therefore follows that for growth to resume at pre-1980 levels, industrial energy use will have to increase over time. For conventionally defined labor productivity to actually increase (as opposed to current labor productivity increases that are achieved by decreasing the denominator), energy use per unit of labor input must increase (energy deepening).

For this to take place, cheap energy must be available (hydrogen?). In short, the conditions that led to the massive energy deepening witnessed in the post-WWII period must be recreated. Cheap energy will, in turn, contribute to the development of new, more energy-intensive capital equipment (technology). Instead of being on increasing second-law efficiency, the emphasis, as far as new technologies are concerned, will be on increasing energy use (consumption).

Zero-Sum Society

Barring a massive increase in the rate of growth of energy consumption, per-capita material wealth, measured by GDP per capita, will remain relatively stable, as it has over the past three decades, over which material wealth has remained relatively constant. This, I maintain, has had and will continue to have a profound effect on social dynamics. Whereas previously the income pie would increase in size, providing larger slices for all, now the pie will no longer grow. Cast in game-theoretic terms, Western industrialized societies can be viewed as zero-sum games, as opposed to non-zero sum games. As such, one's gain(s) will be achieved at the expense of someone else. Capital's gains will be achieved at the expense of labor. Unlike the pre-energy crises era when the income pie was growing, today, social and economic relationships are dialectic in nature. This has manifested itself in a number of ways, including increased income inequality, increased resistance to globalization, and increased social unrest.

unpleasant chapter !

Conclusions

pompous

The events of the post-energy crisis period have, in general, been studied separately. Until now, no one has attempted to provide an integrated view of globalization, automation, the ICT revolution, increasing income inequity, and the failure of Western industrialized economies to return to pre-energy crisis growth levels. In this paper, such an attempt was made. Using the theory of energy rents developed in Beaudreau (1998), the events of the post-energy crisis period were studied and more importantly, integrated. The ICT revolution and off-shore relocations were shown to be flip sides of the same coin, the coin being profit-maximizing responses to the end of energy deepening, and zero energy rent growth. What had been a non-zero sum game, as far as income and wealth were concerned, became a zero-sum game.

The results, we believe, shed new light on the past two decades, decades that can be described as somewhat tumultuous, especially in comparison to the post-WWII period. Optimism, in so far as income and wealth was concerned, gave way to pessimism. Government expenditure, the defining feature of the post-WWII period, gave way to government restraint, not to mention budgetary austerity, deficits, and skyrocketing debt. Centuries, if not millennia, ago, such wrath would have been attributed to the gods, to Apollo, Cautha, Shamesh, Dhatar, and Knich Kakmo.

References

Incomplete Bibliog

Ayres, R.U. and Nair, I. (1984), 'Thermodynamics and Economics', *Physics Today*, pp. 62-71.

Babbage, C. (1832), *The Economy of Machinery and Manufacturing.* C. Knight, London.

Beaudreau, B.C. (1995), 'The Impact of Electric Power on Productivity: The Case of US Manufacturing 1958-1984', *Energy Economics*, 17(3), pp. 231-236.

Beaudreau, B.C. (1998), *Energy and Organization: Growth and Distribution Reexamined.* Greenwood Press, Westport, CT.

Beaudreau, B.C. (1999), *Energy and the Rise and Fall of Political Economy*. Greenwood Press, Westport, CT.

Beiser, A. (1983), *Modern Technical Physics*. The Benjamin/Cummings Publishing Company, Menlo Park, CA.

Berndt, E. and Wood, D.O. (1975), 'Technology, Prices and the Derived Demand for Energy', *The Review of Economics and Statistics*, pp. 259-268.

Blanchflower, D.G., Oswald, A.J. and Sanfey, P. (1996), 'Wages, Profits and Rent Sharing', *Quarterly Journal of Economics*, 60(1), pp. 227-251.

Bresnahan, T. and Trajtenberg, M. (1992), 'General Purpose Technologies: Engines of Growth?', National Bureau of Economic Research Working Paper No. 4148, August.

David, P.A. (1990), 'The Dynamo and the Computer: An Historical Perspective on the Modern Productivity Paradox', *American Economic Review, Papers and Proceedings*, pp. 355-361.

Fuller, B. (1981), *Critical Path.* St-Martin's Press, New York, NY.

Jevons, W.S. (1865), *The Coal Question.* MacMillan and Co, London.

Jorgenson, D.W. (2001), 'Information Technology and the US Economy', *American Economic Review*, 91(1), pp. 1-32.

Lloyd George, D. (1924), *Coal and Power*, London: Hodder and Stoughton.

Maddison, A. (1987), 'Growth and Slowdown in Advanced Capitalist Economies: Techiques of Quantitative Assessment, *Journal of Economic Literature*, 25(2), pp. 649-698.

Soddy, F. (1924), *Cartesian Economics, The Bearing of Physical Sciences upon State Stewardship*. Hendersons, London.

Solow, R.M. (1974), 'The Economics of Resources or the Resources of Economics', *American Economic Review*, 64(2), pp. 1-14.

Solow, R.M. (1994), 'Perspectives on Growth Theory', *Journal of Economic Perspectives*, 8, pp. 45-54.

Temple, J. (1999), 'The New Growth Evidence', *Journal of Economic Literature*, 37(1), pp. 112-156.

United Nations (1984), *Industrial Statistics Yearbook 1984*, New York, United Nations.

US Department of Commerce (1975), *Historical Statistics of the US: Colonial Times to 1970, Bicentennial Edition*. Bureau of the Census, Washington, DC.

US Department of Commerce (various years), *Annual Survey of Manufactures*. Bureau of the Census, Washington, DC.

US Department of Commerce (1986), *Survey of Current Business*, Washington, DC: Bureau of Economic Analysis, January.

Van Reenen, J. (1996), 'The Creation and Capture of Rents: Wages and Innovation in a Panel of UK Companies', *Quarterly Journal of Economics*, 61(1), pp. 195-226.

Wilson, E.O. (1998), Consilience, the Unity of Knowledge. Vintage Books, New York, NY.

Chapter 6

Liberalization of Electricity Markets in Selected European Countries

Paul Welfens and Martin Keim

Economics, Germany

Considering the sustained output growth in OECD countries and high growth rates in Asia and some newly industrializing Latin American countries at the turn of the 21st century, long-term growth of energy demand seems a likely expectation. The supply side of the energy sector is complex and politically sensitive in the case of oil and gas. In regard to oil reserves, the Middle East will continue to represent the world's largest reserves; as for natural gas, Russia and Kazakhstan represent the dominant share of global reserves. Oil will remain the prime energy source for mobility for many decades, while gas together with coal and nuclear fuel plus renewables will be the most crucial inputs for the generation of electricity. Electricity is a vital input factor for all industries and is used by almost all households, which therefore makes its generation, transmission and distribution a sensitive issue. While reducing costs and prices of electricity is desirable from the user point of view, it is also important to guarantee a continuous power supply with close to 100 percent certainty. In advanced industrialized countries, even short-term black-outs can have disastrous effects, as was shown by blackouts in North America in 2003 and in Italy in early 2004. *good backdrop.*

Liberalization of electricity markets has been under consideration in many OECD countries and in some other countries as well. In the case of the EU, liberalization and economic integration – the creation of a single electricity market – go hand in hand; thus naturally both pricing issues and safety of supply have an international dimension. Since the electricity sector is very capital intensive, firms emphasize the need for long-term planning. At the same time, liberalization often goes along with privatisation, which means that formerly state-owned electricity firms become subject to short-term pressure from stock markets. In some cases, this leads to fraudulent management behaviour as in the case of Enron in the US in the late 1990s. *Liberation*

Taking a more long-term view seems to have become a widespread concern among politicians who often emphasize sustainability. However, while this development sounds like the beginning of a new political culture, there are some indications suggesting that sustainability is largely a buzz word. In fact, the decision horizon of politicians did not become more long term in the late 20th century, rather it is getting biased more towards the short term. We will return to this issue in the final section of our analysis, when we examine policy implications.

Liberalisation ⟹ short-termism
Sust'y ⟹ long-termism } TENSIONS
Sust'y as (buzz word)

The EU progressively liberalized electricity markets after 1999; indeed, at face value by the end of 2003, a single electricity market had been established. This represents remarkable progress, since electricity is not only an input into nearly all products and services but is also a politically sensitive area for the following reasons:

- First, it was a monopoly market in many EU countries for decades;
- Secondly, there are universal service obligations, which require the electricity sector to provide access to the network and sell power to any user in the respective country wishing to have access as well as power;
- Thirdly, electricity is an important element of the energy sector, which is largely responsible – together with the manufacturing and transportation industries – for both CO_2 emissions and the global warming problem.

Electricity generation, transmission and distribution are capital intensive. Therefore, investment decisions are facilitated by a stable long-term policy framework. Due to the Kyoto Protocol, there is no clear long-term framework on a global scale. The EU, however, has decided that CO_2 emission trading will be adopted, and this will concern all industries and the energy sector. Emission trading involves an element of uncertainty particularly because the price range of CO_2 emissions traded will not be clear before trading starts in 2005. Such uncertainty is, however, to some extent a normal part of entrepreneurial life; facing risky prospects in decision-making is a natural element of electricity markets since liberalization has begun in the EU. In regard to renewable energy sources, the EU wants to increase the percentage of renewable energy in the total energy supply to 12 percent by 2010, which amounts to doubling it in one decade. In a technical sense, liberalization means, on the one hand, freedom of investment and free market entry, while on the other it signifies freedom of choice on the side of electricity users. The electricity market consists of three layers:

- Electricity generation;
- Electricity transmission (high voltage grid);
- Electricity distribution to firms and households.

In some countries, there are vertically integrated markets – the EDF in France may serve as an extreme example. This is moreover a state-owned company, which – assuming the company is able to utilize its good political connections – is particularly sheltered from potential and actual competition. The UK and Poland are counter examples since both countries have separate grid companies. Germany is in an interim position since it has several major regional electricity companies, which are both electricity producers and owners of the regional transmission network. The grid serves as an essential facility since it stands between power generation and electricity use. Hence access to this grid is crucial for competition. If there is competitive pricing of the grid, electricity prices will be relatively high, while profits for firms owning the grid will be above normal. This in turn encourages X-inefficiency (employment of more labour and capital than is really necessary) and distorts foreign markets. Firms with high profits will have an advantage in acquiring foreign firms, yet it is not clear that the relatively most efficient and innovative firms will expand internationally. Weak competition in the electricity sector could also undermine static and dynamic competition in the generator equipment industry, which might contribute to a lower rate of technological progress and therefore lower growth. Hence lack of competition in one EU country implies:

ESI as difficult economic entity

- Price distortions in the home market;
- Distorted structural change as sectors using larger amounts of electricity will gradually be relocated to foreign countries with lower electricity prices;
- Distort competition abroad since foreign direct investment is distorted;
- Weakened Schumpeterian competition in the generator equipment industry.

Weak competition in energy markets implies relatively high prices and a low rate of innovation, including service innovations associated with the provision of electricity (e.g. benchmarking of electricity efficiency in the case of a multi-plant firm). Lack of innovation in turn undermines the goal of improving energy efficiency in the EU. From this perspective, more competition in the EU is crucial, and particular caution must be observed in regard to moves toward vertical integration and discrimination against firms seeking access to transmission networks. The following analysis looks at key developments in electricity liberalization in selected EU countries – including the upstream gas market. In the last part, we draw some conclusions for further consideration.

Competition necessary for innovation – (but can also reduce innovation)

EU Single Electricity Market

The EU monitors competition in the single market (European Commission, 2001) and has developed a broad set of indicators. As a basic approach for the indicator framework developed, the electricity market is divided into two sub-areas:

- Competitive market areas where focus is on competition in power generation, the role of wholesale markets, and competition in customer supply;
- Non-competitive market areas where the focus is on network access conditions, interconnectedness of the national network, and the influence of regulation.

This methodology is somewhat doubtful as it does not consider key aspects of upstream links (e.g. mergers of electricity and gas companies or mergers between coal and electricity companies), as well as downstream links. Moreover, EU monitoring so far does not consider whether large electricity firms are quoted on the stock market and thus subject to the discipline of stock trading. One also might consider to which extent there is state ownership in the electricity sector. State ownership certainly will undermine non-discriminatory regulatory policies, as close ties between the ministry of finance – responsible for state-owned firms – and the ministry responsible for organizing regulatory policies raise doubts about any promise of non-discriminatory regulation.

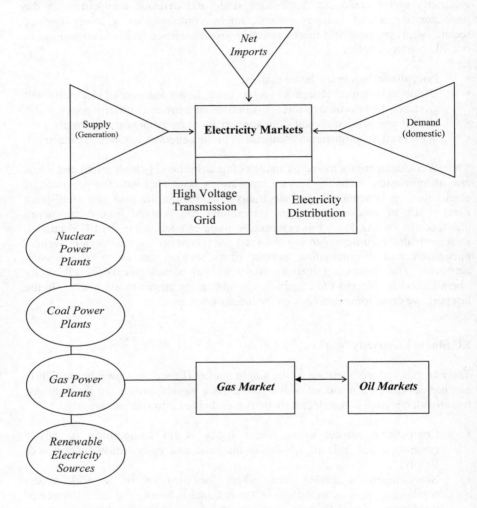

Figure 6.1 Interdependencies in the electricity and energy markets

Energy markets are interdependent. The oil market, shaped by the OPEC cartel with its market power, dominates the gas market, which in turn strongly influences electricity prices, not in the least in spot markets, as gas-powered electricity generation is relatively flexible in its response to changes in demand. In a more general sense, liberalization of the electricity market should not be analysed in an isolated way since upstream markets – e.g., the gas market (with gas being an input factor in gas-powered generators) and the renewables market – are not fully liberalized or are distorted by external effects. Figure 6.1 shows some interdependencies which play a role for competition as well as for merger control.

The Dynamics of Liberalization

As of 1999, the EU has phased in competition in electricity. Generation has been gradually liberalized. Since February 1999, any producer is able to build new power generation capacities anywhere in the Community. Member countries are allowed to apply a tendering approach or an authorization system, through which new plants can be built and operated under the widely applied latter system, if they comply with the planning and energy supply criteria for authorization defined by the respective member country.

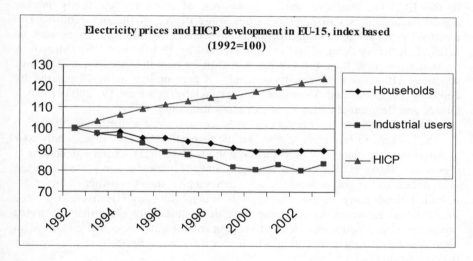

Figure 6.2 Electricity price index for industrial users, private households and the harmonized index of consumer prices for EU-15

Source: Eurostat, own calculations

Note: This table presents electricity prices charged to final domestic consumers, which are defined as follows: annual consumption of 3500 kWh among which 1300 kWh overnight (standard dwelling of 90m²).

The lack of competition in electricity is clearly visible in large price differentials across the 25 EU countries. If there were competition, prices across countries would reflect transportation costs and potentially user preferences with

[handwritten note: lack of competition + price differentials]

respect to the energy mix underlying the generation of electricity. Private households (consumers of 3300 kWh) faced net prices prior to liberalization of between roughly 6 cents per kWh in Greece, Denmark and Sweden versus 12.31 in Germany, 12.45 in Portugal and 16.54 in Italy in 1997. Industrial consumers – with consumption of 50 GWh – faced prices in the range of roughly 3 cents in Finland and Sweden and 4 cents in Denmark, Greece and France, while the highest prices were in the UK, Austria and Germany at about 6 cents (European Commission, 2004). Taking 1996 as the starting point in our price analysis, we can see that the relative price of electricity for both industrial users and private households fell in the period from 1996 to 2003 in EU-15. Relative to 1996, the price index for industrial users fell more strongly in certain time periods than for households.

EU accession countries became active members of the EU single energy market in 2004. On December 19, 1996, the EU countries adopted Community Directive 96/92/EC, which will – with a phasing-in approach – allow for large- and eventually medium-sized purchasers of electricity to choose freely among suppliers in the EU. The directive sets a framework of minimum electricity market regulation and allows member countries some degree of manoeuvrability. The directive provided a gradual liberalization scheme, as follows: 26 percent of national electricity demand had to be liberalized by 19 February 1999. Moreover, consumers of more than 100 GWh p.a. had to be permitted to choose their own suppliers. Of national electricity demand, 28 percent had to be liberalized by February 19, 2000, and 33 percent of demand by February 19, 2003. Ireland, Greece and Belgium obtained some extra transition time.

Access to electricity networks is most critical, since a firm from city/region A, which wants to buy electricity – and the respective services (e.g., a half-hour monitoring of all electricity user sites of a multi-plant firm) – from a generator in region or country C, can get the electricity only if it can be transported over the grid. Access to the grid network by any other supplier than the owner of the grid – so-called third party access (TPA) – is crucial for the effectiveness of the liberalization network. At the same time, it is absolutely clear that integrated producers, who produce electricity in a region and own the respective grid, have no interest in facilitating competition. The higher the grid user price is set, the weaker the competition.

EU liberalization should formally put an end to national and regional monopolies; this could help in recovering all kinds of costs under the heading of state-regulated electricity tariffs. Ideally three effects could be expected:

- One-time-only fall of prices as a consequence of the regime switch from monopoly prices to competitive prices;
- Elimination of X-inefficiencies (i.e., static efficiency criteria will be met);
- Stimulation of process and product innovations in the electricity sector; product innovations could particularly refer to providing an individually desired energy input mix for electricity generation and all kinds of services with a focus on optimal electricity management.

Norway and Switzerland have access to the EU electricity market by way of WTO membership. If Russia joined the WTO, it too would obtain full access to the electricity market in EU-25. Post-socialist transition countries, which even a decade after transition showed energy intensities five to seven times higher than the average for the Euro zone countries, can to some extent be considered natural electricity exporters in the medium term. Once further restructuring and modernization has brought energy intensity closer to the figures for Western Europe, there should also be excess capacity in electricity generation in Eastern Europe and Russia as well as in the Ukraine and Belarus. There is, however, a crucial distinction between EU accession countries from Eastern Europe and Russia. Electricity generators from accession countries will gain access to Western Europe's power grid under the single market framework, while Russia (and other CIS countries) will not. With the EU's eastern enlargement, a trade diversion in goods and (traditional) services could occur with respect to Russia, and likewise an interruption in Russia's electricity exports to eastern Europe.

Electricity Market Liberalization in Scandinavia

After the first European electricity markets began their liberalization processes at the end of the 1980s/early 1990s (England and Wales), Norway became the first Scandinavian country to follow suit (1990). Year after year the other Scandinavian countries started to set market power elements in this market with the exception of Iceland. This section gives a brief overview of the most important steps in the liberalization processes of each country, and shows which 'Nordic' elements are working nowadays to help unify the various markets to a common single market with increasingly strong market elements. Taking a current look at the Scandinavian countries (Iceland will be excluded), we will examine the progress of the processes and their impacts on the market participants; finally we will show some of the differences in the development of each market.

Liberalization of the electricity market in Norway took place via the Norwegian Energy Act in 1990, which came into force in January 1991. The main objectives of the energy act were economic efficiency, security of supply and national equalization of electricity prices. In 2001, about 178 companies were engaged in grid management and operations (42 grid operators, 136 companies are generating and/or trading electricity). Each grid company could own a local, regional or central grid. The state-owned 87 percent of the central transmission grid, through Statnett SF, which is also the operator of the entire central grid. The regional and local grids are owned by municipalities and the counties. Statnett is also the Norwegian transmission system operator and therefore co-ordinates the operation of the entire Norwegian power supply system. It is also responsible for the Norwegian balance system since the relations between electricity production and consumption rates differ quite enormously from region to region. Norwegian electricity generation is largely hydropower (99 percent) while a small remainder is generated by wind power plants. The state-owned Statkraft is the largest producer, with a market share of about 40 percent of all hydropower plants in 2001. Although ten large companies dominate the nuclear power plant market, Statkraft is also the dominant player in this market as well, with a share of nearly 66 percent through its indirect ownership in several companies – in 7 of these 10 companies, Statkraft has the controlling position and is the operator. In 2001, Statkraft's market share totalled approximately 37 percent (NCA, 2002).

SWEDEN   Sweden started the liberalization reform in 1996 and electricity prices were fully liberalized in November 1999. Only the national grid is still a regulated monopoly. The Swedish Energy Agency is responsible for efficiency, reasonable prices, and for a non-stifling sales behaviour of grid companies. Approximately 200 network companies own these grids. Svenska Kraftnät applies a spot tariff on the national grid, which means that all customers who are connected to the grid have access to the entire electricity market and hence can choose the supplier. Svenska Kraftnät is also responsible for the state-owned central transmission network, and for maintaining the balance between production and consumption in all parts of the country as well. The Swedish electricity market is dominated by six companies (Vattenfall, Sydkraft, Birka Energi, Fortum Kraft, Skellefteå Kraft and Graninge) which together have a market share of approximately 93 percent (in 2001). The 11 nuclear power plants are owned by Vattenfall and Sydkraft (these two companies are again dominating the market), and Mellansvensk Kraftgrupp and Fortum. During the whole liberalization process there were many mergers, mainly due to purchases of smaller companies by larger ones. Vattenfall was the biggest electricity producer in 2001 with a market share of 47 percent, followed by Sydkraft (20 percent) and Fortum (17 percent) (NCA, 2002).

FINLAND   A comparable Finnish Electricity Market Act came into force in 1995, but a full opening to all Finnish electricity users did not take place until November 1998. Electricity generation is based on hydropower, nuclear power and mainly on conventional thermal power (mainly bio fuels, coal, natural gas, and peat as well as fuel oil). Finland is dependent on imports from its neighbouring countries since consumption is higher than production. Fingrid owns the national grid and is responsible for the whole national system and for international relations. There are more than 30 balance provider companies. Finland has two nuclear power stations and four, soon to be five, nuclear power reactors (NCA, 2002). In Finland about 120 companies and about 4000 power stations are competing. The five major producing companies and their market shares (altogether about 80 percent) are: Fortum with a market share of 33 percent, followed by Teollisuuden Voima (21 percent), Pohjolan Voima (11 percent), Helsingin Energia (8 percent) and Kemijoki (6 percent).

DENMARK In 1999, Denmark started its liberalization process gradually as the last of the four relevant Scandinavian countries, but since January 2003 consumers are able to choose their electricity suppliers freely. The Danish Energy Regulatory Authority (DERA) has supported these processes since January 2000. The Danish electricity market is divided into two separate markets – Denmark West and Denmark East – which are not interconnected; interconnections merely exist between them and Germany, Norway and Sweden. There are only two, non-interconnected transmission system operators (TSOs) – Eltra and Elkraft. About 100 grid companies are owned by customers or municipalities and have small shares in one of the two TSOs. None of them has a great influence in any TSO. This situation may change with the interconnection between Eltra and Elkraft System in 2004. The stations are Storebælt and The Great Belt. Transmission capacity will be 300 MW (Nordel, 2001).

The Danish market consists of three generators, but only two of them (Elsam A/S in Denmark West and Energi E2 A/S in Denmark East) have emerged as the dominant forces in the market for the time being, through mergers of many existing companies that took place before the Danish Competition Act in 2000. The total installed capacity of Elsam is 7000 MW and of Energi E2 is 5500 MW. The Danish thermal production, which comprised 87 percent of the total Danish electricity generation in 2002, is mainly based on coal and gas. The rest of the whole electricity generation is based on renewable power. As Denmark produces more electricity than it consumes and there is a lack of production capacity in Norway and Sweden, Danish exports become more and more important (NCA, 2002).

Common Nordic Institutions

The Nordic Working group is a common working group consisting of the competition authorities of the five Scandinavian countries (Norway, Sweden, Finland, Denmark and Iceland) which came into force in the beginning of the 1990s. The first European electricity markets started to introduce market power (competition) elements which involved the integration of national markets. As the Icelandic electricity market is not yet liberalized it will not be considered in this section. Since the national markets are regulated by national competition authorities, and the four Scandinavian countries have access to a common wholesale power market, a common working group was established in September 2002 with the following tasks and competencies:

- To identify common Nordic competition issues in the market for electrical power;
- To consider actions to handle obstacles to competition;
- To consider suggestions to amend regulations in order to improve competition;
- to suggest co-operative solutions to improve the effectiveness of competition law enforcement;
- To consider obstacles to competition as a consequence of the integration of actors between different levels of the power market (NCA, 2002).

Nord Pool, the Nordic Power Exchange, was established in 1996 when the Norwegian company, Statnett Marked AS, started a cooperation with the Swedish electricity market and became the first multinational power exchange in the world (Nord Pool ASA). Finland joined the spot market area in 1998, Denmark West in 1999 and Denmark East in 2000. The spot market for physical contracts and next-day-deliveries (day-ahead-auction) are traded on the Elspot market; the largest players are Norway and Sweden (Nord Pool, 2002; Nordic Competition Authorities, 2003): 'Elspot is a marketplace where electricity and capacity is

combined into one simultaneous auction and in cases of bottlenecks different area
prices are established'. Nord Pool also has a financial derivatives market where
future and option contracts are traded, and it additionally offers clearing services
for contracts traded in OTC bilateral contracts. Besides Elspot being traded on the
Elbas (power adjustment) market, physical contracts are also traded, but only
between Swedish and Finnish market participants. 'Due to the lengthy time span of
up to 36 hours between Elspot price-fixing and delivery, participants need market
access in the intervening hours to improve their balance of physical contracts'. The
significance of Elspot has been increasing continuously as turnovers have been rising
steadily. In 2002, physical contracts with a volume of 123.6 TWh have been traded
on Elspot, which is about one-third of the whole electricity generation/consumption
in the Nordic countries. The country shares are: Norway 44.7 percent, Sweden 30.1
percent, Finland 12.0 percent and Denmark 13.2 percent. The trading volume on
the Elbas market was only 0.8 TWh. On the financial market, the volume of the
traded financial contracts was 1,019 TWh (which represents 33.8 percent of the
total volume of all financial power contracts and a value of NOK 180 million ≈
EUR 23 mill.). Future contracts accounted for 54.1 percent of the contracts traded,
40.1 percent were forward contracts and 0.7 percent options contracts (Nord Pool,
2002).

Table 6.1   Key figures of the Scandinavian electricity market, 2002

		Nordel	Norway	Sweden	Finland	Denmark	Iceland
Population	mill	24.3	4.5	8.9	5.2	5.4	0.3
Total consumption	TWh	397.1	120.9	148.7	83.9	35.2	8.4
Maximum load	GW	59.2	17.3	23.3	11.6	6.1	1.0
Electricity generation	TWh	391.6	130.6	143.4	71.9	37.3	8.4
Generation surplus	TWh	-5.5	9.7	-5.3	-12.0	1.9	0.0
Generation surplus	%	-1.4	7.4	-3.7	-16.7	5.1	0.0
Structural Breakdown of Electricity Generation:							
Hydropower	%	55	99	46	15	0	83
Nuclear power	%	22	0	46	30	0	0
Other thermal power	%	21	1	8	55	87	0
Other renewable power	%	2	0	0	0	13	17

Source: Nordel [2002], own calculations

TSOs

Nordel is an organization of the five transmission system operators of the Nordic countries (except Iceland); it includes Statnett (Norway), Svenska Kraftnät (Sweden), Fingrid (Finland), Eltra and Elkraft System (Denmark), and it is responsible for the development of an efficient Nordic electricity market. The cooperation between those TSOs and the most important market players is a key point which must be better understood. Nordel is committed to:

- Act as one Nordic TSO and constitute the basis for a harmonised Nordic electricity market;
- Take a leading role in the development of the Nordic electricity market;
- Furnish leadership also in the development of the European electricity market;
- React quickly to challenges, make decisions and share responsibility for implementing them (Nordel, 2002).

Some figures about the Scandinavian electricity market should give an impression of which countries are producing what kind of power and which countries are dependent on imports or exports of power. Norway and Denmark are interested in boosting electricity exports while Finland and Sweden are dependent on imports. Norway's electricity generation, remarkably enough, is totally based on hydropower (the same for Iceland), while Denmark's is mainly based on gas and coal. Generation in Sweden and Finland is distributed among different types of power.

Market Concentration

The 15 largest companies produce about 81 percent of the total electricity produced in Scandinavia (NCA). It is interesting to see that only three companies have a market share higher than 10 percent but lower than 20 percent (Vattenfall has a market share of 19 percent). The market share of Fortum is very high (16 percent) since both the Swedish and Finnish market are very active. All in all, in the Scandinavian electricity market the Swedish electrical companies have a leading position. The other large companies are Statkraft and Sydkraft (with market shares respectively of 12 percent and 8 percent). The NCA uses the Herfindahl-Hirschman-Index (HHI) to measure the concentration of the Nordic electricity market. The results are:

not that concentrated

Table 6.2   Electricity market concentration (2002)

	Norway	Sweden	Finland	Denmark	The Nordic Market
HHI	1634	2893	1766	4844	892
HHI*	2735	2988	3005	4844	1138

Source: NCA [2002]

The full effects of cross-ownerships are included in HHI (excluding taxes and network tariffs, to different customer categories). The scope is from 0 (an atomistic market) to 10000 (monopoly). The US merger guidelines (United States Department of Justice and Federal Trade Commission, 2002) stipulate an a priori assumption that markets with an HHI below 1000 are unconcentrated, those with an HHI between 1000 and 1800 are moderately concentrated, and with an HHI above 1800 are highly concentrated (NCA, 2002). The tables and figures above show that, under real conditions, each of the four markets is highly concentrated, while the whole Nordic market can be regarded as moderately concentrated.

Price Developments

Generally, prices for electricity vary between customer categories and between urban and rural areas. The final price of electricity consists of the price for the energy, of a network tariff and of taxes (VAT and energy taxes). In the first year of liberalization (1996), the spot price rose by the end of the year. The impacts of increasing competition lowered prices by the year 2000, but in 2001 prices rose again. Although they fell again slightly in 2002, prices remained higher compared to previous years (NEA, 2002). It must be kept in mind that only 40 percent of the final price goes to pay for energy: 20 percent is for network charges and 40 percent is taken up by taxes. Since the system price is highly dependent on variations in hydropower generation, it is reasonable to assume that a high availability of hydropower in the years from 1997 to 2001 led to lower prices than those seen during years of shortage in 2002 and 2003. The high price level in December 2002 reflected pessimistic expectations of further possible scarcities. The price level for network charges remained almost unchanged between 1997 and 2002, but tax levels were continuously increased. This led to higher total costs for end consumers as will be shown in the following table. Note that the development of Swedish prices is quite comparable to the price level on Nord Pool (Elspot) in the same period (SEA, 2003).

Table 6.3   Developments in prices of electrical energy between 1997 and 2003, medium values, Swedish öre/kWh

	Jan 97	Jan 98	Jan 99	Jan 00	Jan 01	Jan 02	Jan 03
Apartment	29.2	29.0	27.1	25.8	27.0	35.6	51.9
Commercial operations	25.8	24.5	23.3	21.0	22.1	28.8	43.6
Small industrial plant	25.6	24.1	22.8	20.4	22.0	28.5	44.3
Medium-sized industrial plant	24.4	23.1	21.6	19.6	21.7	28.3	44.8
Electricity-intensive industrial plant	23.7	22.7	22.5	19.7	22.6	28.3	48.0
Average	25.7	24.7	23.5	21.3	23.1	29.9	46.5

Source: SEA [2003], own calculations

Problems and Prospects of Liberalization in Germany and Austria

Austria fully liberalized its electricity market in 2001. Disregarding taxes (as well as tax changes), electricity prices fell by roughly one-fifth in the period from 1996 to 2003 (E-Control, 2004). If tax increases are included, the result is more modest. In nominal terms, overall prices have increased slightly, but in real terms, prices fell by 2.4 percent. Austria's electricity prices for households were much lower than in Germany in 2003. At 9.2 cents per kWh, the price of electricity in Austria was nearly 1/3 below that of Germany. Interestingly, willingness to change electricity companies is apparently rather low in Austria as only 2.6 percent of households actually did this between 2001 and 2003. For Austrian industry, the electricity price was somewhat higher than the EU average in 2003, which points to some further room for manoeuvring in price reductions.

Germany fully liberalized the electricity market in 1998. However, there are clearly distortions in this market as the use of hard coal is heavily subsidized. By contrast, lignite is competitive to a large extent. Government and industries have decided to phase out nuclear reactors. At the same time, the government implicitly subsidizes renewable energy, although OECD statistics do not cover the type of subsidy granted. For example, those companies that use windmills to generate electricity obtain a guaranteed price – much above the market price level. Hence electricity users are explicitly subsidizing the production of wind energy. Access to the electricity grid is a major problem in Germany, as the country traditionally has been characterized by a half-dozen integrated electricity companies. Negotiated third party access was the approach adopted by the German government (i.e., generators and integrated electricity companies had to negotiate access prices). A similar approach was adopted in the gas market, which yielded unsatisfactory results, since access to the pipeline system seems to be rather restricted. After two agreements concerning access to the gas market, there will be a third agreement for 2003/04. The situation could be improved in the electricity sector (and the gas sector) if principles were imposed guaranteeing non-discriminatory access to the grid. As far as price is concerned, a price cap could be useful, as well as an allowance for the cost of an efficient provision of services – these measures have been quite useful in the telecommunications sector. Rate of return regulation, on the other hand, might encourage inefficiencies – in particular capital intensity would be excessive – as studies on the subject have proved (Averch/Johnson, 1962).

The Federal Antimonopoly Office is supposed to combat the abuse of dominant market share, but it is doubtful that it has had much of a sustained impact. This rather sceptical view is not mitigated by the Office's report that it was conducting inquiries into alleged cases of abuse of market power in a dozen cases – after all, State Antimonopoly Offices were looking into some 200 cases in the same period (Böge, 2002a). Regional distribution in Germany is often in the hands of local government. Local companies are owned by the respective municipality, which often uses profits from electricity distribution to cross-subsidize public transportation. This obviously makes selling the regional distribution network a conflict-prone issue. Germany's major electricity companies have invested in access to customers, and local governments eager to cope with high local budget deficits in many cities have decided to sell off regional distribution companies. The Federal Antimonopoly Office has emphasized that it considers this strategy a problem from the standpoint of real competition, since Germany's two leading integrated electricity firms (E.ON and RWE) effectively constitute a duopoly with respect to the market for large electricity customers and distribution companies (Böge, 2002b). A very delicate case in the German energy market concerned the planned merger of E.ON and Ruhrgas – the latter being Germany's leading gas company, which even has a minority stake in Russia's giant Gazprom. Germany's gas pipeline system is mainly owned by Ruhrgas/VNG, and this, combined with long-term contracts for use of pipeline companies of up to two decades, leads to weak access by outsiders to the pipeline network. Adding new compressors can, however, increase the amount of gas that can be pumped through a given pipeline system. Although the Federal Antimonopoly Office had decided against the merger of E.ON and Ruhrgas, the Ministry of Economic Affairs supported the merger through the application of an exceptional escape clause. The merger of E.ON and Ruhrgas was allowed mainly on the grounds that the relevant market is the wider European one in which a merged company, E.ON/Ruhrgas, would be only one of several major players. However, the problem of reduced access to the Ruhrgas pipeline system was not carefully considered. The merger creates a vertically integrated mega-energy producer in Germany. Whether it will give adequate access to its gas pipeline system and the electricity transmission grid is an open question.

In general, Germany has adopted a broad policy of electricity and gas liberalization. However, the initial plan of the government to solve the problem of TPA on the basis of a cooperative contract framework within the industry has not worked. The almost prohibitive transmission pricing has caused competition in the electricity sector to remain rather weak. The situation in the gas sector is not much better, as there is a duopoly situation which includes pipeline networks. Expansion of gas-powered electricity generation has been anticipated in Germany and some EU countries following earlier examples in the UK and the Netherlands, also because Germany has promised to cut greenhouse emissions drastically in accordance with EU goals and the Kyoto protocol.

Regulatory Aspects in the Electricity Sector

The extent and quality of electricity transmission service can be determined by means of adequate parameters of quality and quantity. The transmitted electric power is a pure quantitative dimension, while quality parameters are the risk and the expected duration of disconnection, the constancy of frequency, and the stability of voltage. Such qualitative features can be summarized in the notion of quality of supply (see Schweppe et al., 1987). These attributes of quantity and quality can only be fulfilled if the requirement of an adequate capacity of the grid is met; more specifically this means the ability of each and every single circuit to transmit its share of desired electricity power at every moment. While instantly transmitted power varies considerably over time, the transmission capacity is fixed in the short run. The short-term cost burden of the grid operator arising from system operation and maintenance services rises by leaps and bounds if the limit of transmission capacity is reached. In the long run expenditures for securing the quality of supply have to be added to the costs of energy losses and maintenance of the grid. These expenditures are caused by the expansion of the grid. In the case of no occurrence of transmission shortages, long-term marginal costs are equal to the short-term ones.

In the case of transmission capacity becoming suddenly insufficient, rationing procedures (e.g. disconnection or refusal of both connection and transmission) must be applied if no market mechanism equalizes the differences between electricity transport service supply and demand. A mechanism like this has to ensure that the grid is first available to users with the highest readiness of payment. The more expensive the use of the grid, the more will users try to reserve and look for alternatives (e.g. bypass). Optimal transmission capacity is reached since marginal readiness of payment of the users meets marginal cost of alternative action in case of congestion, and since the marginal cost of grid scarcity equals the marginal costs of its elimination. In the short run, grid owners can increase their profits by keeping grid capacity limited due to increasing readiness of payment.

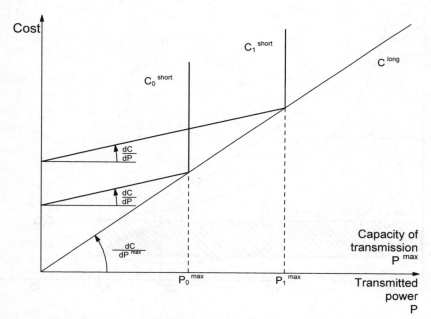

Figure 6.3 Short- and long-term costs of the electricity grid

Source: Kauffmann (2001); see also Laffont/Tirole (1993)

Announcement effects to the users of a grid take for granted that shares of short term and long run marginal costs at total marginal costs are known. For this reason, a subdivision of power transmission services in partial services like physical connection, reserve of transmission capacity and power transmission, scaled for different time zones and individually priced (e.g. for reserve of capacity and transmission, for use of the grid during peak or low load periods), is recommended. Every distinguishable partial service of power grids is limited in terms of quantity and quality; its production requires the employment of limited factors. Efficient allocation of these goods and factors cannot be described simply on the basis of the price level of these bundles of services and factors but implies an optimal price structure. Let us assume first that the grid company is a one-product firm operating the power transmission grid natural monopoly. Due to high fixed costs, the grid company has a U-shaped marginal cost function with average cost C/Q exceeding marginal cost dC/dQ.

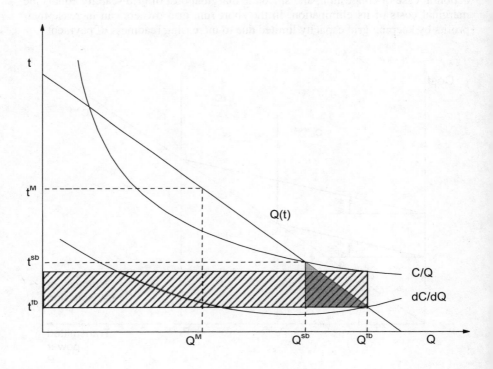

Figure 6.4 Non-coverage of costs due to marginal cost pricing and welfare loss of second best outcome

Source: Kauffmann (2001)

If the grid company priced tariff t for power transmission Q(t) at the welfare optimizing first best level t^{fb} at marginal costs dC/dQ, the company would record losses to the extent of the hatched area in Fig. 6.4. Since a state subsidy to the grid company is not optimal for several reasons (see e.g. Sherman, 1989), a feasible solution to this problem could be price discrimination of a second degree (different prices for different services). For the one-product firm considered now, the best feasible solution is the so-called second best outcome by pricing the tariff at the average cost level C/Q. To reach the second best outcome, the grid company must be allowed to mark up the difference between average costs and marginal costs, resulting in the lowering of power transmission Q and the public welfare loss to the extent of the shaded triangle in Figure 6.4. This kind of regulatory regime, by monitoring the rate of return (or rate of output, sales, or costs, respectively) of the grid company, was applied in practice all over the world for many decades with the strange side effect of overinvestment in grid capacities or/and other facilities of the companies owning the transmission grids. This effect, formalized by Averch and Johnson (1962), is caused by information asymmetries between the grid company and regulatory authorities. To reach more efficient regulatory outcomes, both diversification of services and strategic behaviour of the grid company, connected with asymmetric information, have to be considered. *unnecessarily bignned or pricing*

If we consider the grid company as a multi product firm, the question of the 'right' tariff ratios for different services of the company arises. This question is strongly related to the question of the 'right' taxation rates of a value added tax to be raised on a bundle of goods facing the different price elasticities of every single good. This task was solved by Ramsey (1927). For the regulation of power transmission tariffs, the so-called Ramsey rule implies different treatment of tariffs for services on the basis of their price elasticity of demand. Tariffs for services with inelastic price-demand function should to be more varied than tariffs for services with high price-demand elasticities. This requirement can come into contradiction with our ideas of social justice, because price-inelastic services are often identical with basic human needs. Price cap regulation clearly has its advantages in comparison with price setting based on reported rate of return or cost for the firm (see Beesley/Littlechild, 1989; Vogelsang, 1998). Consider the following:

not v. accessible to the general reader

- Reference to future development;
- High powered incentives for increase of productivity;
- High flexibility of prices;
- Possibility of convergence to RAMSEY prices;
- Easy use by regulatory authority;
- Equalization of short- and long-term marginal costs of the grid.

Disadvantages could arise as the result of a badly chosen value for X, perhaps caused by misevaluation of future developments. As far as distributional aspects are concerned, the introduction of a price cap regulation could become problematic if it leads to huge profits reaped by grid companies, on the one hand, and deterioration of customer groups (e.g. in areas with small population density) on the other. It is one task of the regulatory authority to set clear guidelines regarding the quality of the supply of power transmission. In Germany, electric power tariffs of household customers were traditionally subject to regulation by state authorities at the Ministry of Economic Affairs, which set prices based on Regulation-of-Cost methods. The disadvantages of methods like overinvestment and lack of incentives for cost reduction are well known and frequently discussed (for an overview see e.g. Sherman, 1989, and Borrmann/Finsinger, 1999). The future regulation of power transmission tariffs should be based on a broad set of internal and external information, leading into price caps for bundles of partial transmission services for partial customer groups.

Policy Conclusions

It is clear that a collapse of the OPEC cartel would have enormous implications. The oil price under global competition would hardly be much higher than 10-15 \$, which is less than half of the price fetched by OPEC in the three decades after 1974, the year of the first oil price shock. A fall of oil prices would bring about a fall of gas prices which in turn would raise the demand for gas-powered electricity generation. However, given the fact that the Middle East is the source of most of the world's known oil reserves, and taking into account high long-term growth in oil and gas demand in Asia, including China, one should expect that the relative market share of OPEC will not decline quickly, thereby ensuring its sustained market power for many decades to come.

Upon examination of the electricity market, we find that liberalization has brought only modest gains in Germany and other EU core countries, while the situation in the Scandinavian countries has proved more dynamic. The electricity liberalization results are thus relatively modest in the overall EU. Electricity prices still have not converged much, and there are considerable doubts that transmission fees reflect long-term marginal costs. Indeed, there are two long-term inefficiencies:

- Grid operators and the distribution companies have a strong incentive to levy all fixed costs on the transmission fees; this implies that distribution companies or grid companies that also are power generators will largely impair the access of 'outsiders', in particular foreign suppliers;
- There is also no incentive for firms to restructure existing networks in an efficiency-enhancing way. This in turn reinforces competition among generators, as reduced network charges imply a larger market radius for every power generator that can sell power profitably over the grid.

In many countries, politicians are not eager to promote comprehensive liberalization since the electricity sector is a field in which part of the political elite finds a cosy future environment for retiring from political life. Structural separation will improve competition only after 2007, when distribution companies will no longer be able to charge higher network charges to outsiders than to the own units producing electricity. An unsolved problem is the distortion emerging from CO_2

emissions in countries with a high share of CO_2 intensive energy inputs in electricity generation. There will be an incentive to import more nuclear electricity from abroad where producers often are not facing the full costs of production – e.g. due to artificial limits on liability and thus a bias towards low costs of insurance (France and Germany for example).

The German approach of negotiated third party access is strange, unique in the EU-15 and not efficiency enhancing. In fact, negotiated third party access might increase the risk of collusion in the electricity sector. With the introduction of a new regulator for the electricity sector in 2004, the German system might gradually move towards a more efficient system of resource allocation in electricity. It remains to be seen which approach to regulation the new authority (the existing body in charge of telecommunications and the postal system will be enlarged) will adopt. The special German TPA approach – initially shared by Greece and Denmark – suggests that Germany's political culture has a preference for bargaining approaches and mistrusts market dynamics, which are often perceived as containing a high degree of uncertainty. In contrast, most economists consider competitive markets efficiency enhancing and useful, particularly in bringing about international efficiency gains through trade. At the same time, they mistrust the mechanisms of bargaining among selected players, particularly in the area of political bargaining, which often has a tendency towards short-term perspectives and national solutions.

A rather paradoxical element of energy policy in many EU countries is that it lacks a solid long-term orientation, which would be useful for achieving sustainable development in Europe. Instead, changes in government typically bring major shifts in energy policies, which make the highly capital intensive energy sector – including the electricity sector – a highly politicized business. Foreign investors not well connected to political circles in particular countries and the European Commission might face a crucial disadvantage in the industry. In this respect, US investors are only slightly better positioned than potential investors from Russia or Japan, and this effectively restricts investment from those countries.

In a period of major budget problems in Germany, France and Italy, the government is likely to exploit efficiency gains and price cuts in the electricity sector by imposing new taxes, as was done in Germany and some other EU countries after liberalization. Scandinavian countries are clearly advanced in terms of competition in electricity, and all firms using electricity intensively will particularly benefit from this.

In the EU-15, the political drive towards comprehensive liberalization in the energy sector is undermined by high unemployment in Germany, France and Italy. Introducing competition is likely to lead to lay-offs in part of the industry, and in countries where unemployment already is high, there clearly is resistance to liberalization policies. The electricity sector is capital intensive and characterized by a high concentration in most EU countries so that it is rather easy for well-organized workers and a handful of firms to appropriate economic rents. Germany, France and Italy have a history of resisting reforms and can hardly be expected to quickly move towards comprehensive liberalization. The industry, facing pressures of CO_2 trading, also broadly resists the idea of comprehensive liberalization.

[handwritten margin notes: lower rates of / consumer switching]

In the early liberalization stage, the very low rate of households changing the electricity supply company in EU-15 indicates that establishing a competitive electricity market is a complex challenge. Attempts to create markets for heterogeneous electricity – reflecting the type of input used (so that users can have access for example to 'green power' from renewables at a slightly higher cost) – have largely failed in the EU. *[handwritten: green power hasn't taken off]*

The slow growth of technological progress and the true state of weak competition in the electricity sector is finally revealed by the failure of almost all companies in major EU countries to install metering with remote controls. Only in the UK could one explain annual visits by the meter man from the local distributor with a solid reason, that being the respect for tradition that is an appealing aspect of the British way of life.

References

Averch, H. and Johnson, L.L. (1962), 'Behavior of the firm under regulatory constraint', *The American Economic Review* 52, 1052-69.
Beesley, M.E. and Littlechild, S.C. (1989), 'The regulation of privatized monopolies in the United Kingdom', *Rand Journal of Economics* 20, 454-72.
Borrmann, J. and Finsinger, J. (1999), *Markt und Regulierung*, Vahlen, Munich.
Böge, U. (2002a), 'Building Energy and Transport Infrastructures for Tomorrow's Europe', paper presented at 2. Jahrestagung für Energie und Verkehr, Barcelona, 12 November.
Böge, U. (2002b), 'Liberalisierung der Energiemärkte', paper presented at the Energy Institute of the Free University of Berlin, 18 March.
European Commission (2001), *Electricity Liberalisation Indicators in Europe*, Report by Oxera et al., Brussels.
European Commission (2004), *Third benchmarking report on the implementation of the internal electricity and gas market*, DG Tren Draft Working Paper, Brussels.
Eurostat, http://europa.eu.int/comm/eurostat/.
Kauffmann, A. (2001), *Regulierungskonzepte für den Netzbereich des Strommarktes*. Mimeo, New York.
NCA (2003), *A Powerful Competition Policy, Towards a more coherent competition policy in the Nordic market for electric power; Report from the Nordic competition authorities*, No. 1/2003; Copenhagen, Oslo, Stockholm.
NEA (2002), *Energy in Sweden 2002*, National Energy Agency (Sweden).
Nordel (2001), *Nordel Annual Report 2001*.
Nordel (2002), *Nordel Annual Report 2002*.
Nordpool (2002), *Annual Report 2002*, Nord Pool Spot AS.
Nordpool (2004), http://www.nordpool.no/.
Ramsey, F.P. (1927), 'A contribution to the theory of taxation', *Economic Journal* 37, 47-61.
Schweppe, F.C., Caramanis, M.C., Tabors, R.D. and Bohn, R.E. (1987), *Spot pricing of Electricity*. Kluwer Academic Publishers, Boston.
SEA (2003), *The Electricity Market 2003*, Swedish Energy Agency, Eskilstuna.
Sherman, R. (1989), *The regulation of monopoly*, Cambridge University Press, Cambridge.
Vogelsang, I. (1998), 'Optimal price regulation for natural and legal monopolies', Paper prepared for CIDE seminar on the structural reform and regulation in the energy sector. http://www.cre.gob.mx/english/publications/researchindx.html.

Appendix 6.1 Eurostat electricity retail prices (current prices, before taxes)

	Jul 97	Jul 98	Jul 99	Jul 00	Jul 01	Jul 02	Jul 03	% change since 1/1999	% change since 7/2002
Germany	162	163	158	134	133	126	134	-17%	6%
Ireland	135	127	126	126	126	127	128	2%	1%
Luxemburg	136	137	137	131	121	122	127	-8%	4%
Belgium	146	149	148	146	128	130	122	-18%	-6%
Italy	119	114	115	128	78	101	104	-8%	3%
Portugal	118	115	105	104	105	100	101	-3%	1%
Spain	109	100	98	98	98	99	95	-3%	-4%
EU-15	105	104	102	98	92	92	93	-9%	2%
Greece	84	82	86	83	87	87	90	5%	4%
Austria	160	161	162	126	102	97	89	-45%	-8%
France	91	89	87	85	85	86	83	-7%	-4%
UK	105	105	108	101	93	84	78	-28%	-7%
Finland	59	59	55	54	54	57	68	21%	19%
Denmark	51	52	52	55	65	67	65	23%	-3%
Sweden	69	67	59	53	41	36	46	-26%	30%
Netherlands	91	92	94	101	106				

Source: European Commission (2004)

Notes: Eurostat category Ib; Consumption of 50MWh/year. Prices in the table exclude VAT and other energy taxes.

PART 3:
ENERGY AND SCIENCE

Chapter 7

Science and Education

Petra Lietz and Dieter Kotte

More academic style, many refs.

Social Researchers, Germany

Qualified scientists are vital for the further development of renewable energy sources. Physicists and engineers are needed, for example, to enhance the field of rotor aerodynamics of wind turbines (ECN 2004), chemists are in demand to develop further efficient and cost-effective applications of hydrogen fuel cells (Deluga, Salge, Schmidt and Verykios 2004), biologists are necessary to advance knowledge regarding the use of biomass for primary energy sources (Hoffert et al. 2002) and architects are needed (Pahl 2003) to explore different aspects of natural home heating.

Yet, community labour force data provided by the International Labour Organization (ILO) (2004) show that scientists and engineers constitute only a small proportion of the workforce in the EU, ranging from as little as 2.1 percent in Austria to as much as 8.2 percent in Finland.   *shortage of scientists?*

This shortage of scientists in general, and women in the area of science and technology in particular, has been an ongoing concern to educators and educational researchers (Hodgson 2000; Matyas 1984; Zewotir 1999).

Mau (2003), for example, found in a study of 8th graders that students who were male, and had higher academic proficiency and self-efficacy in mathematics, were more like to persist in their career aspirations in the area of science and engineering. Also in a study of middle school students, Koszalka (2002) found that in addition to gender, more informed and more regular use of Web resources was an important predictor of the desire to pursue a science-related career. Moreover, Gates (2002) has emphasized the importance of male role models – particularly their fathers – for the decision of girls to enter non-traditional areas of work such as engineering.   *lack of role models; misconceptions*

Munro and Elsom (2000) showed that the decision of school leavers to pursue a science-related career was influenced by experiences in science classrooms, science teachers, extracurricular activities initiated by science departments and information about the content of post-16 courses and strategies for coping with advanced studies. Finally, Medhat (2003) and Koszalka (2002) suggested that students' misconceptions with respect to the work of scientists constitute the major reason for students' lack of interest in the pursuit of science-related careers.

Based on these findings, a number of programmes have been designed to increase students' interest in science-related careers. Thus, Chowning (2002) advocates science fairs that bring together high schools, universities, local biotechnology and biomedical companies as well as research institutions in order to facilitate students' explorations of work-related opportunities in the natural sciences. Other initiatives including work experiences during school terms (Moore and Holmes 2003) and summer internships in industry and research institutions (Summer opportunities for students 2003) are aimed at addressing possible misconceptions by providing students with real-life experiences with the work of scientists. The programme put forward by Hammrich and Richardson (2000) is aimed at increasing teachers' awareness of their own role in students' decision-making processes and encourages teachers to promote deliberately science literacy to male and female students alike.

Furthermore, instructional units containing biographies and experiences of scientists provide insights into the actual life and work of natural scientists and are aimed at eliminating the off-putting image of a scientist as a lonely nerd who stares into a microscope all day. Other approaches to teaching natural sciences at the secondary school level have started to include units that explore career and workplace-related issues (CyberEd 1999).

In addition to programmes addressing all school students, many programmes are aimed at increasing particularly the interest of female students in pursuing science-related careers, since the proportion of women in science is still extremely low (Hodgon 2000). Thus, many programmes in colleges are aimed at encouraging women to start as science majors and persevere with their choice (Fisler, Young and Hein 2000; Packard 2003). Once in the workforce, women face frequent obstacles when they try to enter the right networks that could help promote their careers. Here, Davis (2001) discusses especially designed mentoring programmes intended to assist women in science-related occupations.

Yet, despite all these efforts, many countries are still unable to motivate a sufficient number of young people to take up science-related occupations. Sufficiently high levels of people choosing scientific careers would enable these countries to be at the cutting edge of research and discovery in the sciences in general, and in the area of renewable energy in particular.

The present essay tries to examine which variables are linked to a greater interest in science and to arrive at a profile of those students who, at the end of their compulsory schooling, are interested in pursuing a science-related career.

Data

The main aim of this study is to identify variables relating to students' interest in science-oriented careers, particularly towards the end of their compulsory schooling, when they are beginning to consider different educational and occupational possibilities. For this purpose, data obtained by the OECD Programme for International Student Achievement (PISA 2000) were considered appropriate.

The PISA target population referred to those grades in which 15-year-old students were typically enrolled; and the sample in each country was designed to be representative of this target population. Table 7.1 lists those 28 OECD countries and four non-OECD countries that participated in the PISA 2000 study together with the number of students and schools in each sample.

Table 7.1 PISA 2000 – Participating countries, number of students and schools in samples

	N students	N schools
OECD countries		
Australia	5,176	231
Austria	4,745	213
Belgium	6,670	216
Canada	29,687	117
Czech Republic	5,365	229
Denmark	4,235	225
Finland	4,864	155
France	4,673	177
Germany	5,073	219
Greece	3,644	157
Hungary	4,887	194
Iceland	3,372	130
Ireland	3,854	139
Italy	4,984	172
Japan	5,256	135
Korea (Republic of)	4,982	146
Luxembourg	3,528	24
Mexico	4,600	183
Netherlands	2,503	100
New Zealand	3,667	153
Norway	4,147	176
Poland	3,654	127
Portugal	4,585	149
Spain	6,214	185
Sweden	4,416	154
Switzerland	6,100	282
United Kingdom	9,340	362
United States	3,846	153
Non-OECD countries		
Brazil	4,893	324
Latvia	3,920	154
Liechtenstein	314	11
Russian Federation	6,701	246

The PISA 2000 study was the first in a three-year testing cycle designed primarily to generate indicators of educational outcomes for OECD member countries. The first cycle in 2000 focused on results in the area of reading, while mathematics and science served as minor components. In addition to achievement data, questionnaires administered to students collected background information such as socio-economic status of the family, extracurricular activities, and attitudes towards school and learning. Furthermore, information such as size, staffing and equipment of schools was obtained through school background questionnaires.

Of particular relevance to the present study was that the student background questionnaire asked students to indicate their expected future occupation (ST40Q01). Responses were coded according to the International Standard Classification of Occupations (ISCO 1988) developed by the International Labour Organisation. Codes considered to reflect interest in pursuit of a science-related career are given in Table 7.2.

Table 7.2   ISCO-88 codes of science-related occupations

ISCO-88 code a)	Occupation
211	Physicists, chemists etc
212	Mathematicians, statisticians etc
213	Computing professionals
214	Architects, engineers etc
220	Life science and health professionals
221	Life science professionals
222	Health professionals

Notes: a) Includes four-digit subcategories, e.g. 2112 'Meteorologist' or 2145 'Mechanical engineers'.

It should be noted that the ISCO-88 classification lists a wide range of occupations dealing with science in one way or another (e.g., refrigerator technicians, electricians, opticians). However, for the present analysis we have defined as science-related occupations those requiring formal education and training at an advanced level. Occupations which required vocational training only were not classified as science-related ones. This differentiation allowed for a clearer distinction between non-science and science-related occupations.

Method

First, a correlation analysis was undertaken to identify those student performance and background variables which related to the students' expressed interests in science-related occupations. For this purpose, a new variable (EXPSCIH) was computed indicating whether or not students intended to pursue a science-related career, with all students for whom one of the codes in Table 7.2 was recorded being assigned a '1', and all others a '0'. With the exception of the performance indicators, which were measured on a continuous scale, all other variables with *correlation analyses*

which the 'interest' variable was correlated were measured on an ordinal scale. Hence, Spearman rank-order correlation coefficients were used as the appropriate measure of association (Thorndike 1997). Correlation analyses were performed on a data set containing all participating countries in order to reveal variables that related to science-career interest in general.

The PISA 2000 data sets made available by the OECD contain several different science scores. These vary slightly according to the computational method used. For the analyses presented here, so-called 'warm estimates' have been used. A warm estimate is a corrected maximum likelihood estimate from an IRT one-parameter model. In PISA, these estimates were computed in a three-step process:

? eh

(i)   First, the item parameters were estimated based on a subsample of 500 students from each of the OECD participating countries. Non-OECD countries did not contribute to the computation of the item parameters;

(ii)   Second, the warm estimates were computed for all students and all schools by anchoring the item parameters of step one;

(iii) These warm estimates were then standardised (mean = 0, stddev = 1) on OECD countries, except Netherlands as this country did not meet the OECD's strict sampling requirements.

For school estimates the same procedure was applied (except that the school subsample was N= 100). The sum of weights for each country was a constant, so that each OECD country contributed equally to the standardization.

Frequency statistics of the variable indicating science-career interest (EXPSCIH) were used to identify countries for which further analyses were to be undertaken. At this stage, it was decided to focus only on member states of the European Union for which data was available both from the OECD PISA 2000 survey and the relevant ISCO statistics. This limited the number of countries to those presented in Table 7.3. This table gives the proportion of 15-year-old students expecting a science-related job as well as the percentage of the workforce actually found in science-related occupations.

Table 7.3 shows marked discrepancies between the number of students expecting to pursue a career in science and the actual proportion of a national workforce occupied in science-related jobs. While students' expectations regarding science careers run lowest in Austria (8.2 percent), these are more than three times higher in Portugal (29.2 percent), where nearly one third of all 15-year-old students envisage such a career. But, obviously, the high expectations have not become a reality so far. Thus, virtually no difference can be observed in the percentage of workforce occupied in science-related jobs between Austria (2.2 percent) and Portugal (2.3 percent).

Finland, in contrast, reports the highest proportion of the workforce occupied in science-related jobs (9.0 percent). Thus, it is of interest to look at these three countries, namely Austria, Finland and Portugal, in greater detail in order to: a) compare students with an interest in a science-related career across countries and b) investigate what might be linked to the differences in those countries between the

3 – worthy comparison in more detail

proportion of students desiring a science-related career and the proportion of the workforce occupied in such jobs.

cluster analysis

Therefore, in a second step, cluster analyses were conducted on those students in Austria, Finland and Portugal who envisage a science-related occupation.

Table 7.3   Proportion of 15-year-old students opting for a future career in science and actual proportion of national workforce employed in a science-related career in the year 2000 for selected European countries (in percent)

	15-year-old students expecting a science-related job [a]	Workforce occupied in science-related jobs [b]
Austria	8.2	2.2
Netherlands	11.3	5.4
France	12.6	4.5
Germany	13.5	5.4
Sweden	14.9	5.7
Finland	19.6	9.0
Belgium	22.7	7.3
Spain	25.4	3.4
Portugal	29.2	2.3

Might include 'non-professional' science jobs

Notes: a) Based on variable indicating interest in science-related job (ST40Q01 for details see text in section on 'Data'; derived from the PISA 2000 data set (var EXPSCIH). b) Data taken from LABORSTA (Labour Statistics Database) of the ILO available online at: http://laborsta.ilo.org/cgi-bin/brokerv8.exe.

Results

Table 7.4 gives an overview of the student-level variables, which – across all countries that took part in PISA 2000 – exhibit a considerable correlation with the expectation to embark on a science-related career (EXPSCIH). The correlates of science-career interest include students' home background, computer usage, schoolwork, attitudes and performance. Thus, students whose fathers have completed higher education and whose parents have higher-level jobs are more likely to express an interest in a science-related career. Likewise, students who learn things more quickly than others and who do not give up as easily as their peers when subjects or materials are difficult express a greater interest in science-related careers. Finally, greater confidence in their abilities and higher performance in reading and science are linked to the interest in a science-related career.

factors behind desire science job

In addition to listing those variables which correlate with science-career interest across all countries that participated in PISA-2000, Table 7.4 also specifies the actual correlation coefficients between those variables and the interest in pursuing a science-related career for Austria, Finland and Portugal.

Table 7.4 Student level variables correlating with interest in science-related career

Variables measuring	Variables correlating with interest in science-career across all countries participating in PISA-2000	Label	correlation with EXPSCIH [c]		
			AUT[b]	FIN	POR
Home background	Father completed tertiary education (1=yes; 2=no)	ST15Q01	0.09	0.11	0.16
	Socio-economic index of the parents [d]	HISEI	0.13	0.10	0.22
Computer usage	Frequency of reading e-mail and web pages	ST36Q05	0.07	0.13	0.16
School work	Understand difficult material presented in text	CC01Q02	0.08	0.18	0.11
	Understand difficult material presented by teacher	CC01Q08	0.07	0.22	0.12
	I keep working, even if material is difficult	CC01Q12	0.07	0.18	0.13
	I can do an excellent job on assignments and tests	CC01Q18	0.07	0.19	0.06
	I can master skills taught to me	CC01Q26	0.05	0.13	0.10
	Sometimes I get absorbed doing maths	CC02Q01	0.04	0.16	0.13
	Because maths is fun, I don't want to give it up	CC02Q10	0.06	0.25	0.13
	I get good marks in maths	CC02Q12	0.07	0.25	0.13
	Maths is one of my best subjects	CC02Q15	0.05	0.28	0.11
	I've always done well in maths	CC02Q18	0.06	0.26	0.17
	Maths is important to me personally	CC02Q21	0.02	0.27	0.15
	Amount of homework science done each week	ST33Q03	0.01	0.06	0.19
	Number of minutes per week in science courses	SMINS	0.13	n.s.	0.27
Attitudinal Composites	Control expectation (if I decide to learn well/not to get bad grade/not to get problems wrong, I can) [d]	CEXP	0.05	0.19	0.11
	Self-concept (Academic; learn things quickly, do well in tests, good at most subjects) [d]	SCACAD	0.11	0.23	0.12
	Self-efficacy (CC01Q02, CC01Q18&CC01Q26) [d]	SELFEF	0.09	0.20	0.12
Performance	Reading performance score	WLEREAD	0.17	0.15	0.21
	Science performance score	WLESCIE	0.19	0.18	0.22

Notes: a) All student level variables recorded which correlate substantially ($r \geq |0.10|$; p<.000) with interest in a science career (EXPSCIH; coded 0=no, 1=yes) across all countries taking part in PISA 2000. b) AUT=Austria, FIN=Finland, POR=Portugal. c) EXPSCIH=derived variable measuring students' expectation to resume a science-related career; coded as 0=no, 1=yes. d) Derived variable/construct (see Adams and Wu 2002). n.s. correlation coefficient is not significant with p≥0.05.

Substantiated by similar results found across all countries, the home background of a student correlates positively with EXPSCIH in Austria, Finland and Portugal. This result supports previous findings (Keeves 1991, Kotte 1992), since science-related careers are usually demanding and require tertiary level education. This, in turn, is often more readily achieved where parents themselves have obtained tertiary qualifications. This relationship is most noticeable in Portugal (r=0.22), the EU country with the lowest per-capita income of the three (OECD 2003).

Performances in reading and science are also clearly and significantly correlated with the expectation to embark upon a science-related career in each of the three countries. Even stronger correlations emerge for Finland than for Austria and Portugal between science career expectations and the students' interest and attitude towards mathematics. This evidence confirms results of previous research which found that higher achievement was linked to a greater desire to pursue a career in science (Miller, Lietz and Kotte 2002).

Students' perception of their work at school as well as their attitudes – as measured by the PISA composites on the academic self-concept (SCACAD) and self-efficacy (SELFEF) – show the highest correlation coefficients with science-career expectation in Finland and the lowest correlation coefficient in Austria. In contrast, the amount of time students are exposed to science lessons at school (SMINS) does not correlate with EXPSCIH in Finland, whereas it does correlate in Austria (r=0.13) and Portugal (r=0.27). This is particularly noteworthy as the mean of SMINS (measured in minutes per week) differs considerably between the three countries (Austria: 151; Finland: 204; Portugal: 235).

The question, therefore, is the extent to which those students who express an interest in a science-related career might differ across countries. To shed light on this question, cluster analyses of only those students in each country who envisaged a science-related occupation for themselves were conducted separately for Austria, Finland and Portugal (Hair, Anderson, Tatham and Black 1995), using the Statistical Package for the Social Sciences (SPSS). The variables that served as the basis for the cluster formation were those that previously had been identified as being linked to student interest in a science-related career (see Table 7.4).

Two clusters emerge from the analyses in each country. In other words, students with an interest in a science-related career fall into one of two groups. The mean value – also called 'cluster centroid' – for each of the variables that differentiate significantly between the two clusters in Austria, Finland and Portugal is given in Table 7.5.

After an examination of those variables in which cluster centroids differ noticeably, the two clusters have been labeled 'realists' and 'dreamers' respectively. These labels intend to show that the 'realists' are those students whose science-career aspirations appear to be warranted by their science performance. The 'dreamers', in contrast, might aspire to a science-related career, yet their background, attitudes and performance seem to suggest that it is unlikely that they will actually achieve their aspirations.

While some aspects of the clusters are similar across countries, others vary. Below, the results of the cluster analysis are first described separately for each country with an overarching discussion of the findings ensuing in the subsequent section.

Table 7.5   Student level variables and their mean values for the two clusters of students who intend to pursue a science-related career in Austria, Finland and Portugal

Variables measuring	Label	Austria		Finland		Portugal	
		realists	dreamers	realists	dreamers	realists	dreamers
Home background	HISEI	-	-	55.20	50.11	53.23	47.90
School work	CC01Q08	-	-	-	-	2.82	2.64
	CC01Q12	-	-	2.73	2.95	3.07	2.78
	CC02Q01	-	-	2.74	3.00	3.10	2.91
	CC02Q12	-	-	3.22	2.88	-	-
	CC02Q15	-	-	3.06	2.76	-	-
	CC02Q18	-	-	-	-	2.82	2.53
	ST33Q03	1.92	2.59	-	-	3.22	2.66
	SMINS	172.16	273.47	160.36	292.48	580.94	172.41
Attitudinal composites	CEXP	-	-	0.44	0.15	0.03	-0.16
	SCACAD	-	-	0.57	0.15	0.42	0.08
	SELFEF	0.60	0.21	-	-	0.23	0.05
Performance	WLEREAD	586.42	491.68	605.20	515.46	569.84	483.93
	WLESCIE	597.15	506.06	602.16	497.25	549.34	477.72

Notes:

Variable label	Variable explanation
HISEI	Socio-economic index of the parents
CC01Q08	Understand difficult material presented by teacher
CC01Q12	I keep working, even if material is difficult
CC02Q01	Sometimes I get absorbed doing maths
CC02Q12	I get good marks in maths
CC02Q15	Maths is one of my best subjects
CC02Q18	I've always done well in maths
ST33Q03	Amount of homework science done each week
SMINS	Number of minutes per week in science courses
CEXP	Control expectation: If I decide to learn well/not to get bad grade/not to get problems wrong, I can.
SCACAD	Self-concept, academic: I can learn things quickly, do well in tests, good at most subjects.
SELFEF	Self-efficacy: I understand difficult material in text, can do excellent job on assignments and tests, can master skills taught to me
WLEREAD	Reading performance score
WLESCIE	Science performance score

Austria

In Austria, the two clusters are defined by five variables for which the cluster centroids differ significantly. For each variable, Table 7.6 provides information of the significance testing of differences between the cluster centroids. The F-ratio and its probability value indicate that the difference in group means for the two clusters, i.e. 'realists' and 'dreamers', is, in fact, significant.

Thus, remarkable differences exist between clusters when it comes to the time spent on science homework (ST33Q03) and on science lessons at school (SMINS). Even more pronounced is the difference between the two clusters in regard to student performance in science (WLESCIE) and reading (WLEREAD). While the 'dreamers' spend considerably more time on science lessons at school and on homework, their performance lags behind that of the 'realists'.

Table 7.6   ANOVA for student level variables forming the two clusters of 'realists' and 'dreamers' – Austria

Variable	Cluster MS	DF	Error MS	DF	F	Prob
ST33Q03	19.2334	1	0.926	198.0	20.7660	.000
SMINS	440058.5938	1	9355.939	199.0	47.0352	.000
SELFEF	6.9281	1	0.871	201.0	7.9521	.005
WLESCIE	363643.5665	1	4480.180	201.0	81.1672	.000
WLEREAD	393378.9064	1	3882.267	201.0	101.3271	.000

This finding is, indeed, unexpected. Typically, more science exposure is associated with higher performance (Keeves 1991; Kotte 1992; OECD 2001). One possible explanation might stem from the type of instruction Austrian students experience. A lesson in a science-specialized class may be more effective as far as student performance is concerned than some standard or generic science instruction. Clearly more background information and in-depth research would be needed to explain this observation.

The only other variable, next to exposure to science and performance, which discriminates the two clusters is self-efficacy with a clear difference between 'realists' (0.60) and 'dreamers' (0.21) in Austria. This shows that the 'realists' are more confident than the dreamers when it comes to understanding difficult material in a text, doing an excellent job on assignments, and mastering the skills they are being taught.

Finland

As is the case for Austria, Finnish students who express an interest in a science-related career can be grouped into two clusters. Again, the performance level of the 'realist' cluster is considerably higher than that of the 'dreamers', with a difference between the two groups of nearly 105 points for science, and 90 points for reading. Moreover, the former group has to put in less effort to achieve at this higher level as reflected in the noticeably lower amount of time spent on homework in science by the 'realists'.

Unlike Austria, however, the two clusters seem to stem from noticeably different home backgrounds, with the 'dreamers' coming from homes with a lower socio-economic status (HISEI). At the same time, the 'dreamers' might get more absorbed in doing maths (CC02Q01) than the 'realists', but it is the latter group who reports higher performance in maths (CC02Q12, CC02Q15).

Table 7.7 ANOVA for student level variables forming the two clusters of 'realists' and 'dreamers' – Finland

Variable	Cluster MS	DF	Error MS	DF	F	Prob
SMINS	1600300.9713	1	8245.584	426.0	194.0798	.000
CC01Q12	4.5757	1	.526	441.0	8.6946	.003
CEXP	7.7975	1	.825	443.0	9.4426	.002
CC02Q01	6.2594	1	.751	440.0	8.3254	.004
HISEI	2493.6326	1	261.937	438.0	9.5199	.002
SCACAD	17.0286	1	.884	441.0	19.2631	.000
CC02Q12	11.0871	1	.817	441.0	13.5631	.000
CC02Q15	8.6394	1	.995	441.0	8.6830	.003
WLESCIE	1071961.8842	1	4961.553	443.0	216.0537	.000
WLEREAD	787998.8495	1	5360.597	444.0	146.9983	.000

The two clusters of Finnish students who express an interest in a science-related career also differ significantly with respect to their control expectation (CEXP). Thus, the 'realists' are more convinced that they will not get bad grades or problems wrong if they set their mind to not doing so, unlike the 'dreamers'. In addition, the cluster of students which shows higher performance in science and reading also exhibits a higher academic self-concept (0.57), compared with the 'dreamers' (0.15). In total, ten variables distinguish between the two clusters in Finland (see Table 7.7).

Portugal

In Portugal, the cluster profiles are based on the largest number of variables. Here, significant differences emerge for twelve variables between the two clusters (see Table 7.8).

In terms of marked differences in home background (HISEI), attitudes (CEXP, SCACAD, SELFEF) and performance in science and reading (WLESCIE, WLEREAD) the results for Portugal confirm those for Austria and Finland.

However, the Portuguese cluster of 'realists' shows a clear difference when compared with the corresponding cluster in Austria and Finland. In Portugal, the

significant natural differences

'realists' are exposed to more than three times the amount of science instruction (581 min) compared to the cluster of 'dreamers' (172 min). Also, unlike the 'realist' cluster in Austria and Finland, 'realists' in Portugal devote significantly more time to homework in science than do the 'dreamers'. In addition, the mean difference in achievement between the two clusters is smaller in Portugal than it is in Austria or Finland (reading: 86 points; science: 71 points). However, it should be kept in mind that the average science achievement for all students in Portugal was lower than those of the other two countries (OECD 2001).

In Portugal, attitudinal differences between the two clusters are rather pronounced. These differences are consistent with the findings in the other two countries. The academic self-concept is higher for 'realists' (0.42) compared to the group of 'dreamers' (0.08) as is the self-efficacy (0.23 cf. 0.05).

Table 7.8   ANOVA for student level variables forming the two clusters of 'realists' and 'dreamers' – Portugal

Variable	Cluster MS	DF	Error MS	DF	F	Prob
ST33Q03	44.6021	1	.774	624.0	57.5888	.000
SMINS	22648624.6792	1	9770.034	599.0	2318.1724	.000
CC01Q12	13.2108	1	.654	683.0	20.1747	.000
CEXP	5.7936	1	.793	683.0	7.2973	.007
CC02Q01	5.9321	1	.709	679.0	8.3637	.004
HISEI	4326.1307	1	266.005	672.0	16.2633	.000
CC01Q08	5.2822	1	.581	681.0	9.0794	.003
SCACAD	17.9527	1	.873	682.0	20.5432	.000
SELFEF	5.1381	1	.758	683.0	6.7768	.009
WLESCIE	794080.8576	1	6386.622	684.0	124.3350	.000
WLEREAD	1142515.3728	1	5870.714	684.0	194.6127	.000
CC02Q18	12.1445	1	.965	677.0	12.5753	.000

After having identified the composition and discriminating variables for the two clusters in Austria, Finland and Portugal it is useful to examine the size of each cluster. Table 7.9 presents the proportion of the total sample in the three countries represented by each cluster. In addition, the mean test score in science for each of the subgroups is given.

Table 7.9 shows how the proportion of the sample represented by the two clusters varies across the three countries. Thus, the cluster of 'realists' is biggest in Finland with 13.2 percent of the Finnish sample in this cluster. In Austria, this cluster represents only 5.6 percent of the sample. This is an interesting finding when taking into account the percentage of the workforce actually occupied in science related jobs (see Table 7.3, second column). It seems that in a country such as Finland students are more realistic and their desire to pursue a science-related career is rooted in an actual higher performance in science. This, in turn, leads to a larger proportion of students who pursue science-related careers later on in life.

Table 7.9   Proportional cluster size and mean science score of students interested and not interested in a science-related career – Austria, Finland and Portugal

cluster/ subgroup	Austria		Finland		Portugal	
	%	mean science score	%	mean science score	%	mean science score
realists	5.6	597.2	13.2	602.2	10.0	549.3
dreamers	2.6	506.1	6.4	497.3	19.2	477.7
no science career expectation	91.8	504.9	80.4	527.8	70.8	459.1
total	100.0	508.0	100.0	534.0	100.0	468.7

Furthermore, the proportion of the sample represented by the 'dreamers' differs considerably between the three countries (Austria: 2.6 percent; Finland: 6.4 percent; Portugal: 19.2 percent). Thus, in Portugal relatively more 'dreamers' are found than in the other two countries. A possible explanation might stem from Portugal's lower average income compared with Austria and Finland (OECD 2004). As a consequence, Portuguese students 'dream' of a science career, which they associate with a high status, more readily than their counterparts in the other EU countries.

Discussion

This study identified a number of variables associated with the intention of 15-year-old students to pursue a science-related career. Correlative analyses of data collected as part of the PISA-2000 survey showed that a higher level of parental socio-economic status, more frequent use of computer for reading e-mails and web pages, a higher academic self-concept and self-efficacy was reported by those students who expressed an interest in a science-related occupation.

In the next step, the proportion of 15-year-old students who aspire to a science-oriented career was compared with the proportion of the workforce actually employed in science-related occupations. These comparative data were compiled for those European countries that participated in PISA-2000. While Finland had the highest proportion of the workforce pursuing careers in science, Austria reported the lowest one. Moreover, Portugal had the largest difference between the proportion of students manifesting such an interest at age 15 and the proportion of the workforce that actually ends up following a career path in science.

In order to examine the way in which this group of students interested in a science-related career might be different across the three countries, cluster analyses were performed separately for Austria, Finland and Portugal. In each country, two clusters emerged. One cluster was formed by those students who actually did very well in science, i.e. whose performance in PISA 2000 was noticeably higher than the country average of all 15-year-old students.

As the desire for a science-related career appeared to be based on actual high performance in science at school, this cluster was labelled 'realists'. Most of these 'realistic' students were found in Finland (13.2 percent), followed by Portugal (10.0 percent) and Austria (5.6 percent).

If the 'realists' constitute the top group of students, the second cluster could be considered the 'wishful thinkers'. These 'dreamers' also expressed their interest in a science-career but their actual performance was either at or below the national average in each of the three countries.

Further examination of the cluster profiles showed that variables in addition to performance measures differentiated between the 'dreamers' and the 'realists'. Thus, 'realists' came from homes with a higher socio-economic status, felt greater control over their learning, and had a higher academic self-concept than the 'dreamers'. In Austria and Finland, this latter group, in contrast, had a considerably lower science performance than the 'realists' despite spending more time in science classes and doing more homework.

Finally, a closer look at the proportions of the total sample represented by each group, namely those who are not interested in a science-related career, and the two clusters with such an interest revealed cross-country differences. Thus, in Portugal, the fact that a far greater proportion of 15-year-olds could be considered 'dreamers' – in that they aspire to a career in science but their performance in this field is only just above average – might be explained by the lower economic performance of this country in comparison to Austria and Finland.

This finding gives rise to the hypothesis that in countries with lower GDP students' aspirations for certain careers might be based primarily on the desire to attain a higher socio-economic status rather than their actual performance in those subjects necessary to embark on such a career path. Hence, it would be of interest to examine this hypothesis by extending the analyses reported in this article to all countries that participated in PISA 2000 and to compare the proportions of 'dreamers' and 'realists' for countries of higher and lower GDP.

References

Adams R. and Wu, M. (eds) (2002), *Programme for International Student Assessment (PISA): PISA 2000 Technical Report,* OECD, Paris.

Chowning, J.T. (2002), 'The student biotechnology expo: A new model for a science fair', *American Biology Teacher* 64(5), 331-339.

CyberEd Inc. (1999), *Interactive biology* [TM] Multimedia Courseware Series. [CD-ROM]. Retrieved June 15, 2004 from http://www.cybered.net. Chica, CyberEd Inc., CA.

Davis, K.S. (2001), 'Peripheral and subversive, Women making connections and challenging the boundaries of the science community', *Science Education* 85(4), 368-409.

Deluga, G.A., Salge, J.R., Schmidt, L.D. and Verykios, X.E. (2004), 'Renewable hydrogen from ethanol by autothermal reforming', *Science*, *303*(5660), 993-997. Retrieved October 4, 2004, from http://www.sciencemag.org/cgi/content/full/303/5660/993.

Energy Research Centre of the Netherlands (ECN) (2004), *Wind turbine technology*. Retrieved June 15, 2004, from http://www.ecn.nl/wind/research/turbine_technology/index.en.html.

'Finding out about careers in science' (2003), *Education in Science* 203, 12-16.

Fisler, J.L., Young, J.W. and Hein, J.L. (2000), 'Retaining women in the sciences: Evidence from Douglass College's Project SUPER', *Journal of Women and Minorities in Science and Engineering* 6(4), 349-72.

Gates, J.L. (2002, April), *Women's career Influences in traditional and nontraditional fields*. Poster presented at the 9th Biennial Meeting of the Society for Research in Adolescence (New Orleans, LA).

Hair, J.F., Anderson, R.E., Tatham, R.L. and Black, W.C. (1995), *Multivariate data analysis* (4th edn.), Englewood Cliffs, NJ, Prentice Hall.

Hammrich, P.L. and Richardson, G.M. (2000, April/May), *The sisters in science program: Teaching the art of inquiry*. Paper presented at the Annual Meeting of the National Association of Research in Science Teaching (New Orleans, LA).

Hodgson, B. (2000), 'Women in science – Or are they?', *Physics Education* 35(6), 451-53.

Hoffert, M.I., Caldeira, K., Benford, G., Criswell, D.R., Green, C., Herzog, H., Jain, A.K., Kheshgi, H.S., Lackner, K.S., Lewis, J.S., Lightfoot, H.D., Manheimer, W., Mankins, J.C., Mauel, M.E., Perkins, J., Schlesinger, M.E., Volk, R. and Wigley, T.M.L. (2002), 'Advanced technology paths to global climate stability: Energy for a greenhouse planet', *Science* 298(5595), 981-987. Retrieved October 4, 2004 from http://www.sciencemag.org/cgi/content/full/298/5595/981.

International Labour Organisation (2004), *Geneva (workforce data), LABORSTA Labour Statistics Database, 1998-2000*. Retrieved March 12, 2004, from http://laborsta.ilo.org.

International Standard Classification of Occupations – 1988 version (2004), Retrieved September 27, 2004, from http://europa.eu.int/comm/eurostat/ramon.

Kotte, D. (1992), *Gender differences in science achievement in 10 countries – 1970/71 to 1983/84*, Peter Lang, Frankfurt.

Keeves, J.P. (1991), *Changes in science education and achievement, 1970-1984*, Pergamon Press, Oxford.

Koszalka, T.A. (2002), 'Technology resources as a mediating factor in career interest development', *Educational Technology and Society* 5(2), 29-38.

Matyas, M.L. (1984, April), *Science career interests, Attitudes, abilities, and anxiety among secondary school students: The effects of gender, race/ethnicity, and school type/location*. Paper presented at the Annual Meeting of the National Association for Research in Science Teaching (New Orleans, LA).

Mau, W.-C. (2003), 'Factors that influence persistence in science and engineering career aspirations', *Career Development Quarterly* 51(3) 234-43.

Medhat, S. (2003), 'Tapping young potential: Are we investing enough in science, engineering and technology?', *Education in Science* 203, 8-10.

Miller, L., Lietz, P. and Kotte, D. (2002), 'On decreasing gender differences and attitudinal changes: factors influencing Australian and English pupils' choice of a career in science', *Psychology, Evolution and Gender* 4(1), 69-92.

Moore, M.J. and Holmes, W.R. (2003), 'Biology experience impacts career development', *American Biology Teacher* 65(5), 355-59.

Munro, M. and Elsom, D. (2000), 'Choosing science at 16: The influence of science teachers and career advisers on students' decisions about science subjects and science and technology careers', Cambridge, England: National Institute for Careers Education and Counselling.

Organisation for Economic Co-operation and Development (2001), *Knowledge and skills for life. First results from PISA 2000,* OECD, Paris.

Organisation for Economic Co-operation and Development (2004), *Economic outlook: June 2004 No. 75,* OECD, Paris.

Packard, B.W.L. (2003), 'Student Training Promotes Mentoring Awareness and Action', *Career Development Quarterly* 51(4), 335-45.

Pahl, G. (2003), 'Natural home heating: The complete guide to renewable energy options', White River Jct., VT, Chelsea Green Publishing Company.

Programme for International Student Achievement (PISA) (2000), *Data sets.* Retrieved March 10, 2004 from http://pisaweb.acer.edu.au/oecd/oecd_pisa_data_s1.html.

'Summer opportunities for students' (2003), *Winds of Change* 18(1), 31-33.

Thorndike, R.L. (1997), 'Correlational methods', in J.P. Keeves (ed.), *Educational research, methodology, and measurement: An international handbook* (2nd ed., pp. 484-493), Elsevier Science, Oxford.

Zewotir, T. (1999), 'Gender differences in science career choice', paper presented at the Annual Meeting of the Australian Science Teachers Association (48th, Adelaide, South Australia), Retrieved June 15, 2004 from: http://www.science.adelaide.edu.au/sasta/conasra.

Chapter 8

Tomorrow's Scientists:
Where Will We Find Them?

Linda Miller *UK, raial scientist*

The need for human ingenuity in making discoveries and creating new products, services and processes means that the success and innovation is critically dependent on the availability and abilities of scientists and engineers. It is therefore vital that the supply of science and engineering graduates with appropriate skills keeps pace with greater investment in R&D and innovation, and with the demand for these skills ... (Roberts, 2002).

Unclear need – given
v. Low proportion of workforce
in science-based jobs...!

The UK is experiencing a crisis in science.

Recent reports, commissioned by the UK government departments for education and skills and for trade and industry, indicate that there are problems at almost all stages of the recruitment and retention of scientists. Studies have pointed to the problems that England is currently experiencing in encouraging young people to enter science at the university level and in employment (Roberts, 2002); the difficulties in retaining staff in science, engineering and technology (PSP and IER 2002); and the particular difficulties of recruiting and retaining women in science (Greenfield, 2002).

English focus /
women focus

What Types of Scientists?

In the UK there is a tendency to talk in broad terms about 'scientists' or the 'SET' sector – science, engineering and technology. In fact, the number of people entering the sciences increased by around 11 percent between 1995 and 2000, but most of this increase was in biological sciences and computer science. In engineering and technology, however, the number of entrants fell by some 7 percent in the same period, and physics and mathematics entries fell by 1 percent.

falling number of engineers

Perhaps these figures do not sound significant, but first we should remember that this has taken place against a backdrop of expansion in the HE sector and of concerted efforts to encourage wider access to higher education. Secondly, there is an additional problem concealed within the data on biological sciences. Biological science graduates are far more likely to work as teachers after graduating, rather than in research and development, than graduates of the other science disciplines.

Around 55 to 62 percent of graduates in physics, engineering and technology, chemistry and computer science tend to work in R&D, compared to around 30 percent of graduates in biological sciences. Hence the R&D personnel are being mostly recruited from that part of the pool of science graduates that is gradually shrinking.

If we are to develop new technologies, this effort will be crucially affected by the availability of new graduates. In his report to the Government, Roberts makes the following observation:

> A shortage of graduates in these disciplines is likely to become increasingly serious since the UK economy – with its ... strong science base and increasing focus on high-tech and high-added value manufacturing businesses – is likely to need more mathematics and physics graduates (Roberts, 2002).

[margin handwritten note: demand for scientists]

To encourage innovation, we need an expanding and increasingly broad and diverse pool of potential recruits, not one that is dwindling.

Of course, we often hear about the 'brain drain', the migration of UK scientists to the USA, so it might be reasonable at this point to consider whether there is a relatively simple explanation of the problem in the salaries offered in the UK. In fact, market forces appear to be having an impact on salaries in engineering and technology. Roberts cites data from the Labour Force Survey for the UK from 2001 that shows that jobs for graduates in subjects that are in short supply – such as engineering and technology – offer significantly higher salaries than those in the biological or social sciences. So it would not appear that an easy solution can be found by simply offering higher salaries. Market forces have ensured that this has happened already, and it does not appear to have worked.

Therefore, this chapter explores the nature of the current recruitment problem, give a brief overview of research that has been conducted to date, and consider what further actions and developments might be recommended, based on those findings.

At School and University

The supply of entrants to university science courses is threatened by the growing unpopularity of science as a subject for study at school. The Roberts review reported that, between 1991/2 and 1999/2000 the number of entrants for A-levels (the qualification that is typically taken at around age 18, and is the traditional qualification for those entering higher education) in physics and mathematics fell by 21 percent and 9 percent respectively. He suggested that a range of issues were affecting young people's interest in science:

[margin handwritten note: Roberts Review]

- Shortages in the supply of science and maths teachers, along with a lack of appropriately-qualified teaching staff;
- Out of date laboratories and equipment;

[margin handwritten note: as starting point]

- Courses that failed to interest and inspire pupils, combined with poor teaching styles and methods;
- A range of other factors that included the views of teachers and family in relation to science subjects, and the nature of career advice provided to young people.

These last two points will be expanded in the review of research in this area.

Roberts expressed both general concerns regarding the falling numbers of young people entering science, and also raised some specific issues regarding the progress of females in SET. In examinations taken at age 16, girls tend to outperform boys in all subjects except physics. However Roberts (2002) noted that 'the proportion of girls studying mathematics and the physical sciences post-16 is still considerably lower than that of boys', and this is despite the fact that, at A level (age 18), girls outperform boys in all subjects except English, history and computer studies, for which their performance is the same as that of boys.

When we look at entries to university courses, we find that while women constitute more than half of entrants to higher education in England, they constitute a minority of science undergraduates and employed scientists.

Around 30 percent of entrants to all SET courses are female. The situation is worse in engineering and technology, where females comprise around 12-13 percent of intake, and little better in physics, where they comprise around a fifth of entrants. In Chemistry, the number of female undergraduates decreased by 17 percent between 95/96 and 00/01 (although it should be noted that the number of males fell even faster), while in Engineering, the percentage of female undergraduates has remained stable between 95/96 and 00/01 at 14 percent.

So far we have been concentrating on the traditional route from school through university. But when considering vocational education and training, even more extreme figures are found. For vocational qualifications in engineering (Advanced Modern Apprenticeships for those aged up to 25 and National Vocational Qualifications for older trainees), women constitute just 3 percent of current entrants to these awards.

The last section considers the specific difficulty of encouraging more women to enter science. Roberts gave a clear message that 'the under-participation of women in SET is damaging the UK's supply of scientists and engineers' (Roberts, 2002). There is a general acknowledgement that it is only by recruiting from the whole of the potential workforce that the UK will meet its skill needs. In the context of IT, Patricia Hewitt has said: 'If we're only recruiting from half the potential talent pool it's really not surprising that industry faces a skills shortage.'

The rest of the chapter focuses on what might be done to attract more women into science. Given the statistics presented, there is clearly quite a battle ahead in order to turn the situation around. But my argument is that by making science more attractive as a career option to women, we may also make it more attractive to a wider pool of men as well.

Research into Career Choice

Consider some of the factors that have been noted as affecting interest in science. Let us look at factors broadly grouped under two headings: course content, teaching style and method; and attitudes, advice and information.

Course Content, Teaching Style and Method

The Roberts Review commented on the poor quality of much of science teaching in schools in the UK, exacerbated by poor facilities and equipment. It expressed concern that many young people viewed the curriculum as lacking any relevance to real life, and that this lack of relevance was a particular issue for girls. Teaching methods failed to inspire or even to interest young people. He observed that practical problems, demonstrations and active learning techniques should be adopted.

In making these recommendations he was in tune with the findings made by Petra Lietz, Dieter Kotte and myself following our attempts to path model data from TIMSS – the Third International Mathematics and Science Survey (Miller, Lietz and Kotte, 2002). We found that one key influence on the development of favourable attitudes towards both science and jobs in science was the adoption of approaches to teaching such as using real life examples, encouraging pupils to work in small groups together, and the use of demonstrations in class. These interactive teaching approaches impact both directly on the development of attitudes, and also indirectly via the development of favourable attitudes amongst the peer group of friends. The attitudes of the friends to science then, in turn, further influence the child's attitude to science. Effectively, this is a 'virtuous circle' in which good teaching has an impact through a variety of routes. So of course does bad teaching.

More interactive teaching style needed

virtuous + vicious circle

Attitudes

While there is some debate over whether family background and parental occupation is influential or not, what appears less contentious is the influence of parental attitudes in shaping the young persons' attitude toward science. Breakwell and Robinson (2001) looked at changes between 1987-88 and 1997-98 in the attitudes of males and females towards science. They found first, that an interest in science had declined over the period. However, they also found that parental influence, and particularly mothers' perceived support for science was an important factor in developing positive attitudes towards science.

Our work with the TIMSS data set (Miller, Lietz and Kotte, *ibid*) confirmed the importance of the mothers' influence, not just through developing favourable attitudes to science but also by encouraging the belief that increased effort at school will result in greater achievement. It is worth commenting that this derives from what is called the 'attributional model', in which we see that people generally tend to attribute their successes either to effort or to ability. Women are often found to attribute their performance to ability, or more accurately perhaps, to lack of ability. This is a problem, since we are pretty much stuck with our abilities, whereas we are able to decide how much effort we devote to a task. Encouraging young people to believe that increasing the amount of effort they expend on a task can improve their achievement levels is a critical step in improving their achievement rates; what the model produced from the TIMSS data set shows that there is a link first between mothers' attitudes and teaching approach that leads to the development of attributions to effort, then from those increased attributions to achievement level in science tests, and then from the increased achievement levels to more favourable attitudes towards science and careers in science.

This is all well and good when parents can be relied on to instil positive messages about science. But how often is this the case? There are limited routes to influence parental attitudes. The path model however would seem to suggest that adopting better teaching methods in the classroom might go some way towards compensating for any lack of encouragement at home and through this help to promote the development of more favourable ideas about science.

Information and Advice

The Roberts Review observed that people had quite limited and outdated ideas regarding what scientists do. Outdated stereotypes of scientists – usually working alone, almost always male (Haste, 2000) – are reinforced by media representations. These ideas fail to be challenged by science teachers who often do not hold a specialist science qualification. Roberts found that it was not uncommon for teachers at secondary school level to be teaching subjects for which they did not hold a relevant degree, and in some cases did not even hold the relevant A level. Thus stated, it is not surprising that many teachers are ill-equipped to give advice on careers in science. It should be noted that young people view their teachers as primary sources of information about careers (Reid et al., 2003).

The fact that there is little accurate information about the work of scientists would seem to point to the need for accurate information to be provided by career advisers. However, there is evidence that many pupils do not receive up-to-date or accurate advice on science careers from advisers (Munro and Elsom, 2000). Furthermore, there appears to be a view that studying science will reduce the options available to the young person, rather than be a valuable generic qualification (Osbourne and Collins, 2000).

There have been attempts to improve the image and accessibility of science with a range of projects offering 'hands-on' experience to young people. The sorts of things that have been tried include visits by scientists and engineers to schools to talk about science and engineering, and buses equipped as travelling 'laboratories' that visit schools to offer extra-curricular 'hands-on' science and engineering activities.

SEMTA – the authority that sets the requirements for vocational training in science and engineering and oversees training in this sector – commissioned a review of all 'hands-on' activities to determine the extent to which they were encouraging interest amongst girls and improving young people's interest in careers in science and technology (SQW, 2002). What they found, though, was that these projects failed to make any link between the activities and possible jobs, and

(margin note: poor outreach activities)

for those projects that visited schools, the teachers also failed to make any link between the activities and possible careers, or indeed between the activities and the science/technology curriculum. These findings rather reinforce the message from the Roberts Review concerning the inadequacies of much science teaching and career advice in schools at present.

Given this situation within schools, it seems all the more necessary to ensure the availability of accurate and up-to-date career information. However, Munro and Elsom (2000) report that career advisers appear not to challenge young people's career stereotypes and views on scientific subjects. Around 40 percent of career advisers felt that mathematics and science were 'different' in terms of career guidance. The research indicated that there was a lack of systematic training and updating in occupational information available to career advisors. Young people make subject choices at a time when their motivation in science subjects is decreasing and their perceptions of science careers are very hazy. For many pupils, their individual interview with a career adviser came after they had made the decision to drop science.

(margin note: poor careers advice)

The need for information about careers in science is particularly important given that young people are largely motivated only to find out information about jobs once they are persuaded that the job might be of interest to them. In our work with schoolchildren, Rowena Hayward and I demonstrated that there are sex differences in the knowledge of young people regarding various sex-stereotyped jobs, and a correlation between the extent to which young people think that they would like a job, and the amount they claim to know about it. There appeared to be some particular problems with some of the scientific jobs.

Table 8.1   Differences in liking for, and knowledge about, scientific jobs amongst 14-18 year olds

	Boys' liking for job	Boys' knowledge of job	Girls' liking for job	Girls' knowledge of job
Computer engineer	3.28	2.69	2.20	2.31
Civil engineer	3.09	2.31	2.04	1.85
Forensic scientist	2.76	2.54	2.60	2.21
MLSO	2.62	2.43	2.32	2.14
Materials scientist	2.56	2.38	2.10	1.95
Biotechnologist	2.30	1.54	2.08	1.32
Molecular geneticist	2.23	1.59	2.08	1.33
Software engineer	2.15	2.42	2.28	2.13

Scale: 4 = 'yes, I know what this job involves'; 3 = 'a fairly good idea of what this job involves'; 2 = 'some idea of what this job involves'; 1 ='No, I don't know what this job involves'.

Knowledge about these jobs was generally low, but in each case girls' knowledge was significantly lower than that of the boys.

While it is difficult to prove the causality of any such correlation between interest in, and knowledge about the job, it is also difficult to argue against the notion that it would be better to equip young people with more information about jobs, in the hope that this will lead to more interest. It is surely better to do this than to leave them ignorant about career possibilities in one whole sector of the economy.

The value of career advice is emphasised by work by Miller and Petrie (2002) that looked at the factors that influenced males' and females' decisions to enter IT and computer engineering courses. Their findings suggest that while both boys and girls who enter IT and computer engineering courses were alike in their interest in the subject and the feeling that this was what they really wanted to do, what differentiated them, though, appears to be the extent to which they felt they had already made career decisions. The factors rated highly by males as influencing their decision indicated that they believed that the qualification would help them gain employment in a job or within an area that they had already chosen. The factors rated more highly by females indicated that they had reached a firm decision regarding their future career. They believe that the qualification they had chosen would maximize the chance of employment. The most important factors for males were the belief that the qualification was the best to equip them for the area they wanted to enter, they had a specific job or career in mind for which the qualification would be necessary, they had a specific sector in mind and the course would maximize the chances of entering the sector. For females, the most important considerations were that it was a respected qualification, potential employers would view it as the best qualification to have, they had a specific sector in mind and the course would maximize the chance of entering that sector. If they did not yet know which sector they wanted to work in, then they believed that the course would maximize their chances of getting a job.

This suggests that there are small but critical differences in the factors that influence the educational decisions for males and females. Emphasizing slightly different aspects of subject areas may enable more effective promotional material to be designed to market HE courses in these areas to women, with no immediately obvious detriment to the appeal of the courses to males.

Conclusions

This has necessarily been a very brief review of the topic. The focus has been on recruitment and we have not even begun to touch on issues of retention. Further information on the PSP/IER and Greenfield reports are available from the web and details are given below.

All in all, it would appear in England that there is little prospect of any significant improvement in the near future. The Roberts Review made many recommendations, some of which are to be implemented – improved Continuing Professional Development for science teachers, for example. However, other

[handwritten margin notes: "otter, order, order factors, mainly against science", "gloomy"]

factors, such as the fact that science and technology courses often are a year longer than other degree programmes, may become increasingly significant given the recent decision of the government to allow universities to charge students increasing levels of 'top-up fees' for university tuition.

Perhaps only the arrival at a point of real crisis will persuade the Government to adopt some 'joined-up' thinking in regard to science. We appear to have sufficient research to have a reasonable understanding of the factors that, even if they do not actively militate against, certainly do not assist with encouraging the choice of a career in the SET sector. This chapter has discussed some of the issues that particularly serve to inhibit the movement of *women* into science and engineering. However, if the government could bring in a coherent programme – of improvements to both teacher training and the career advice system, perhaps in conjunction with some focused funding aimed at increasing admissions to science and technology courses – then it is likely that these basic improvements would result in improvements to both male and female recruitment. The likelihood of any such 'joined up' programme being produced in the near future however, is doubtful.

Note

This chapter draws on research undertaken by Linda Miller, Petra Lietz and Dieter Kotte while Linda Miller was a visiting scholar at the Faculty of Business and Law at Central Queensland University and we would like to extend our thanks to the Faculty for the funding that made this work possible. In addition, thanks are due to the Equal Opportunities Commission in England for commissioning myself and colleagues at IES to undertake research into gender segregation in engineering, ICT, construction, plumbing and childcare, which contributed towards some of the material in this report. I would also like to thank: Professor Petra Lietz and Nick Jagger for helpful discussions that led to the development of this paper, as well as Emma Pollard and Nick Jagger for allowing me to draw on their review of factors influencing females' choice of careers in science, engineering and technology.

References

Haste, H. (2000), *Prometheus, Pandora and the Sorcerer's Apprentice: Friday Evening Discourse*, The Royal Institution, London.

Miller, L. and Hayward, R. (in preparation), *New jobs, old occupational stereotypes: Gender and jobs in the new economy*.

Miller, L., Lietz, P. and Kotte, D. (2002), 'On decreasing gender differences and attitudinal changes: factors influencing Australian and English pupils' choice of a career in science', *Psychology, Evolution and Gender*, vol. 4 (1), pp. 69-92.

Miller, L. and Petrie, H. (2002), 'Gender segregation in IT: what influences choice of course and career?', paper presented at Gender Research Forum Conference 'The Gender Pay and Productivity Gap', November, London.

Munro, M. and Elsom, D. (2000), *Choosing science at 16: the influences of science teachers and careers advisers on students' decisions about science subjects and science and technology careers*, CRAC, Cambridge.

Osbourne, J. and Collins, S. (2000), *Pupils' and parents' view of the Schools Science Curriculum*. Kings College London.

Pollard, E., Jagger, N., Perryman, S., Van Gent, M. and Mann, K. (2003), 'Ready SET Go: A review of SET study and careers choices', report to the Engineering Technology Board.

Roberts, G. (2002), *SET for Success: the supply of people with science, technology, engineering and mathematics skills. The Report of Sir Gareth Roberts' Review*, TSO, London.

Reid, A., Martin, S., Denley, P., Clarke, C., Bishop, K. and Dodsworth, J. (2003), 'Tomorrow's World, Today's Reality – STM teachers' perceptions, views and approaches', a report for the Engineering Technology Board.

SQW (2002), 'Review of girls-only hands-on experience opportunities in STEM', a Segal Quince Wickstead report for SEMTA.

Other Material

European Commission (2003) *She Figures: Women and Science Statistics and Indicators*. EC 2003.

Greenfield, S., Peters, J., Lane, N., Rees, T. and Samuels, G. (2002), 'SET Fair: A report on women in science, engineering and technology' from The Baroness Greenfield to The Secretary of State for Trade and Industry.

PSP and IER (2001), *Maximising returns to science, engineering and technology careers*, DTI, London.

PART 4:
ENERGY AND LIFESTYLE

Chapter 9

Obstacles to the Use of Renewable Energy Sources in Bulgaria

Antoaneta Yotova

Meteorologist , Bulgaria

Poorly written + structured

The rational use of natural resources, energy resources in particular, has become a matter of great interest in both economic and political agendas since the second half of the last century. Environmental considerations have been added because of the serious concerns about the impact on the environment due to the economic development of modern societies. The energy-economy-environment complex is now being examined in policy studies, in regard to related policies, strategies, programmes, plans, etc. In the energy policy field, the renewables as an option that meets, to the greatest extent, the requirements for environmentally compatible and sustainable energy development, receives more and more attention on global, regional, national and local scales. This is true in Bulgaria, too.

According to the present Bulgarian Law of Energy (LoE), in force since the end of 2003, 'renewable energy sources are ... solar, wind, hydro and geothermal energy, waste heat, energy from vegetable or animal biomass, including biogas, energy from industrial and municipal solid waste which are renewing without visible exhaustion when used with definite power'. This definition of renewables includes all types of such sources available in the country, but in practice, only hydro energy is used as a primary energy source for electricity production in Bulgaria. The share of hydro energy in the country's electricity generation has *hydro* varied between 7-8 percent in recent years. Comparing the figures for the potential of the existing renewables in the country (Table 9.1) and the Bulgarian and the EU energy policy targets (to double the share of renewables in gross domestic energy consumption in the European Union by 2010 from the present 6 percent to 12 percent [European Commission, 1977]), questions arise such as 'For what reasons are renewables not used to their full potential in the country?', and, 'How can the policy targets can be achieved if the present situation continues?'.

This chapter aims to find answers to such questions in Bulgaria. The general legislative framework for the use of renewables in the country is the basis for analyzing the real situation, in this respect focusing on the obstacles that hamper the faster and more effective penetration of renewables. After a short description of the Bulgarian energy sector and of its present renewables, the obstacles to the use *Intro* of renewables are described and analyzed. There follows a review of the obstacles faced by other countries in applying for membership in the EU. The differences between common and country-specific obstacles are considered. Finally, we analyze the last type of obstacles with particular regard to Bulgaria.

The Energy Sector, the Energy Policy and Renewables in Bulgaria

The energy sector in Bulgaria can be generally described in the following manner. 'Bulgaria has historically followed a very energy intensive development policy. At the same time, its energy resources are very limited' (The 3rd Bulgarian National Communication to the UN Framework Convention on Climate Change, 2002). Fourteen years after the beginning of the transition from a centrally planned market economy, the development of the electricity market in Bulgaria has been modest, and it has gradually started to function only in the last few years. Still, the limited availability of domestic fossil fuels – mostly low-quality lignite coal – and a high dependency on import fuels (about 70 percent of the primary energy sources are imported) are the main factors that frame development of the energy sector in the country. Since 1970, nuclear energy has become a significant factor for the country's energy system and policy, but its role has recently become quite controversial.

Renewables in Bulgaria consist of hydro energy, which especially plays a role in the peak-load hours for the country's energy system, and contributes to the country's electricity generation by about 7-8 percent. Now, the installed capacity based on renewables is 1 146 MW and the total share of renewables in the country's energy balance is 0.4 percent. Here, along with hydro energy, some other renewables are included, mainly biomass (wood, agricultural and forestry wastes) used for household heating and industrial plants in the small towns and villages.

Table 9.1 Technical potential of renewables in Bulgaria

Hydro: Technical potential (TWh)	Solar PV: Technical potential (TWh)	Wind: Land area with wind velocity between 4 and 7m/s (1000 km^2)	Wind: On shore technical potential (TWh)	Biomass: Forestry and Agricultural Wastes (kt/yr)	Biomass: Industrial and Municipal Wastes (without Landfill gas) (kt/yr)
3	2	111	11	38 823	3 343

Source: Office for Official Publications of the European Communities, 1994

In order to analyse the utilization of renewables, it is essential to know the existing potential of such sources in the country. Because the potential of renewables is relatively uncontroversial and estimates about it are similar within the short-term framework, we use the results from one of the many studies on the potential of renewables in Bulgaria completed in the last 10 years, namely TERES

(Office for Official Publications of the European Communities, 1994). According to this study, the technical potential for wind, solar photovoltaic (PV) and biomass energy is of considerable value in Bulgaria, as shown in Table 9.1. Also, the potential for large hydro energy plants in Bulgaria is practically exhausted; however, this is not true for the small and micro hydro power plants, nor for biomass, in a short-term perspective (Bulgarensky, 3rd National Conference on Renewables, 2003).

As the next step in the analysis of the use of renewables, it is important to know the main 'actors', i.e. institutions, organizations, etc., responsible and playing active roles in the country's energy system and policy, including matters related to the renewables. Article 3 of the present Bulgarian LoE states that 'the state policy in the energy sector in Bulgaria is defined by the Council of Ministers and implemented by the Minister of Energy', i.e. the Ministry of Energy and Energy Resources (MEER). All activities within the energy sector are being regulated by the State Energy Regulatory Commission, an independent legal body established in 2002. In respect to the policy related to renewables, the Executive Agency on Energy Efficiency (EAEE) and the 'Energy Resources' reporting to the MEER are the governmental bodies responsible for the development and realization of such policy. A number of research organizations, Non-Governmental Organizations (NGOs) and some private companies have become more and more active in this field in the last 10 years, too.

The economic and political conditions of the country, as well as its international relations and obligations, provide the context for the Bulgarian energy policy. The conventions and agreements ratified by the Bulgarian Parliament form the general framework for all related activities. In respect to the role and use of renewables, the UN Framework Convention on Climate Change (UNFCCC) and the Kyoto Protocol, ratified respectively in 1995 and 2002 by the Parliament, are of special significance. An important factor for the promotion of renewables in Bulgaria in the last few years is the process of accession to the EU and, respectively, the harmonization between the Bulgarian and European legislations. The transition towards a market economy, especially the privatisation process that has gathered speed in the last few years, also plays a role, particularly the privatisation of small hydro power plants.

At present, the main instruments for promotion of renewables in Bulgaria are mostly legislative, and include laws (especially the energy law, but also Law on Water, Law on Corporative Income Taxation, etc.), strategies, programmes, action plans, etc. For example, a special chapter in the present LoE deals with the promotion of electricity production from renewables and the issue of Combined Heat and Power, and attention is given to the use of renewables in the new Law on Energy Efficiency recently approved by the Bulgarian Parliament. In Bulgaria, the sub-law normative acts and regulations play an essential role for the practical implementation of laws, but their development is usually a very slow process. To assist in the realization of the legislative framework, and to increase considerably the share of renewables in the national energy balance, a National Programme on Renewables has been developed in the last few years. This Programme, which is coordinated by the EAEE, contains at present nearly 1000 concrete investment projects and project proposals from all administrative regions and public sectors in Bulgaria. The database of this Programme is an open system for new projects and proposals as well as for project package formation. In the beginning of March 2004, the EAEE opened an information centre for consultation on energy issues, including the use of renewables.

Obstacles to the Use of Renewables

General and Specific Obstacles

There are many barriers and obstacles to the penetration and use of renewables worldwide, and the same situation applies in Bulgaria as well. These barriers and obstacles can generally be divided into two main groups: the common and the country specific. We will discuss these obstacles in the course of a review of EU applicant countries (Danyel Reiche (ed.), 2003) in order to distinguish what is common and what is country-specific. The focus will always be the current situation in Bulgaria. The common obstacles are those which, in one way or another, prevent the penetration of renewables in most of the countries. These are usually economic constraints – for example, the very high investment costs of designing and constructing devices for electricity generation from renewables, especially solar PV sources. Also, in a number of countries, including Bulgaria, the price of even conventional energy is high compared to the income of citizens, and that is why the energy produced from renewables is not economically attractive for consumers. For these reasons, electricity from solar, wind or biomass energy sources, in spite of their potential in the country, are not yet included in the Bulgarian energy system, and the existing devices are small scale private or experimental ones.

After reviewing the obstacles to the use of renewables in the applicant countries, it is difficult to distinguish clearly between the common and country specific problems, since the existing literature from those countries describes the local obstacles in different ways (some analyses being very brief, others very detailed.) Generally, the obstacles may be classified as political, structural, instrumental, economic and cognitive. Most often, the following are identified:

- Political: lack of a Green Party or other parties to support the use of renewables with appropriate legislative or other relelvant initiatives or to oppose initiatives that do not favour the use of renewables;
- Structural: the difficulties of including in the grid the electricity generated through the use of renewables are of a technical nature, due to the monopoly of state-owned electric companies;
- Economic: high investment costs, risks and prices of the electricity from renewables;
- Cognitive: minimal or no awareness and interest among the general public in the benefits of renewables. The information available is often insufficient to educate people concerning the current energy situation and change individual attitudes and behaviours.

In each country, these common obstacles may appear in a specific combination. The situation in Bulgaria is discussed below.

Country Specific Obstacles to the Use of Renewables in Bulgaria

The country-specific obstacles that stand in the way of a wider and more rapid penetration of renewables in Bulgaria can be summarized as follows: appropriate laws, programmes, strategies, action plans, etc. already exist, or are under development, in accordance with the country's foreign policy and international obligations, but in most cases they remain on paper and have not been implemented. The problems arise from the political and economic realm and relate to the very close relations with the former Soviet Union. Indeed, the position of Russian companies as major suppliers of primary energy sources, importing oil, natural gas and nuclear fuel in Bulgaria, remains very strong. Again, because of Russian support, the 'nuclear lobby' is very strong also in Bulgaria, and this plays a negative role in the penetration of renewables in the country. Bulgarian nuclear energy is a very controversial issue in negotiations regarding application for membership in the EU due to the requirement for phasing out the first two units of the Kozloduy nuclear power plant in 2004 and the third and fourth units in 2006. However, this does not advance the debate regarding renewables. In fact, a negative factor in the process of joining the EU is that the pre-application funds in Bulgaria are managed only by the Ministry of Finance, which essentially limits the abilities of other institutions, even the governmental ones, to use these funds for specific purposes like the penetration and implementation of renewables. In combination with the common economic constraints like high investment costs, risks and the price of electricity (more than 5 euro cents/kWh since the increase of July 2003), it is quite difficult for renewables to be competitive in Bulgaria. In particular, the facts are as follows:

- The actions and measures for the promotion of renewables foreseen as an expression of political will in the present LoE are being put into practice very slowly, if at all. Often, the update of an old law or the drafting of a new one begins before the previous law is entirely implemented. As stated in the chapter for energy of the Regular Report on the progress of Bulgaria in the application to the EU in 2002 (Bulgarian Ministry of Energy and Energy Resources), an '(...) active and co-ordinated policy for an increase of support to ... utilize renewables is still to be defined. The EAEE has no clear mandate, and its governance is not yet settled because of the recent change of the institutional framework and needs support as an element of a wider policy ...'. According to the report for 2003, 'progress in ... the use of renewables is limited'. The necessary changes in other laws have not been made in sufficient time to contribute to the diffusion of renewables in the country (for example, a lack of land consolidation prevents the widespread use of renewables in the agricultural sector).
- Economic structures, especially market ones, are not yet fully developed in Bulgaria, above all in the energy sector, and for activities promoting

renewables in particular. The parallel processes of market economy development and economic restructuring presents special difficulties for the diffusion of renewables. The present high interest rates for bank loans (14-18 percent), as well as lack of real incentives, such as preferential low interest loans to both producers and consumers, do not contribute to a wider utilization of renewables.

- Relevant institutional and infrastructural changes in the Bulgarian energy sector are not occurring fast enough to ensure the necessary conditions for actual adoption and implementation of the best EU and world practices concerning the use of renewables.

- Consumer awareness of the immediate and long term advantages of using renewables is very low in Bulgaria. Also, there are few attempts to promote renewables by targeting individual consumers or offering practical support to small and medium enterprises involved in renewable-related activities. Intelligible consumer-oriented information, such as green electricity standards and labels, 'tips', instructions, specific guidelines on creating projects for the utilization of renewables, etc., are entirely lacking.

- The social and cultural factors have not been considered adequately, if at all, either in research or in energy policy implementation in Bulgaria. This is true for the whole period during which the economy was centrally planned, i.e. from 1944 to 1989. After 1990, progress in economic and policy terms can be characterized as good. In fact, a number of new and modern energy laws, instruments, measures, etc., were developed and partly introduced. Nonetheless, practically nothing was done to evaluate and, where necessary, change the people's understanding, attitude and behaviour regarding energy use in general and, in particular, the implementation of renewables. Having almost totally ignored the social and cultural factors has, in turn, led to very few practical results in the use of renewables in the country.

What Is To Be Done, and How?

As mentioned before, international agreements and obligations like the UNFCCC and the Kyoto Protocol play an essentially positive role for the diffusion of renewables. However, the many problems that the international community has recently faced in negotiations to implement these agreements, especially those related to the Kyoto Protocol, make it more difficult to exert a practical influence on the process of expanding the use of renewables in certain countries. On the other hand, although the requirements of the Kyoto Protocol were not obligatory until February 2005, the recent EU environmental and energy policies have been almost entirely based on its targets (reduction of greenhouse gas emissions by 8 percent in the period 2008-2012 in comparison with the 1990 levels). Bulgaria's intention to join the EU in 2007, imposes on the country the same commitments and targets, and this can be a good basis for the diffusion of renewables. The goal of a sustainable future development also plays a positive role.

Renewable Energy + Energy Efficiency = Sustainable Energy

The UNEP Sustainable Energy Finance Initiative states: 'Instead of climate change from continued investment in fossil fuels, we need to create a climate for change towards a sustainable energy path'. Steps in this direction are:

*   A learning and networking event is planned for the international finance community entitled, 'Creating the climate for change' – to help decision makers in the finance sector get a better grip on the challenges and opportunities associated with investing in clean energy.
*   The International Conference for Renewable Energies (Bonn, Germany, 1-4 June 2004).
*   The second world renewable energy forum entitled 'Renewing civilization by renewable energy' organized by the World Council for Renewable Energy (also in Bonn, from 29 to 31 May 2004).

So although generally, it is generally known what to do, the question remains how to proceed within the concrete conditions and circumstances of a country like Bulgaria. Even the best legislative background for renewables is not enough, if there is no understanding within the sphere of politics and a real willingness to undertake practical actions on the sources faster and wider implementation. Given the situation, it is necessary to focus on people and examine the way they perceive the problem and understand it, and see whether they are willing to take action towards its solution. Most people are unable to grasp the complex issues concerning global warming, and when it comes to renewables, they cannot understand their role in decreasing GHG emissions, which are one of the main factors contributing to anthropogenic climate change in the last few centuries. It is worth making serious efforts in this direction, i.e. to 'translate' relevant research results and legislative measures into understandable explanations, guides and 'tips', so that laymen have a real chance to introduce RES in their everyday lives at an affordable price. Scientists can play an essential role in changing the culture surrounding renewables. By communicating research risults in a way that provides the general public with timely, understandable, practical, and useful information, they can halp shape its attitudes, behaviours and lifestyle. First of all, in order to push for a more intensive use of renewables, the public has to know about the 'external costs' (i.e., the impact on human health, ecosystems and materials) of the production and consumption of all energy sources used in the country. Unfortunately the number of publications on this subject in Bulgarian is still very limited. Therefore the experts, working together with well-informed journalists, should look for special ways to bring this useful knowledge to lay people. The principle behind these strategies is that the accumulation of the actions of single but persistent individuals can be very effective. This approach does not require large-scale investments. Specific initiatives in this direction are:

*   At the individual level: to use the power of the Internet. For example: the International Climate Symbol should appear on the most popular web sites,

in all relevant media, and should be used by all concerned scientists and experts in their communications. The EUGENE, European Green Electricity Network, should become widely known, and as many 'green labels' as possible should be certified by the competent organizations in each country. The modules developed for distance learning and education should be advertised and disseminated in order to become known to and used by as many students as possible.

- At the collective level: to initiate and realize, or take part in relevant projects, like the recent one on 'Bulgarian national capacity self-assessment for global environmental management', funded by the Global Environmental Facility under the auspices of the Ministry of Environment and Water.
- At the community level: to join and be active in initiatives like (1) 'Science cafes' by EUROSCIENCE aiming at '... sharing knowledge ... (and) shaping a critical assessment of societal issues by increasing people's awareness of the basic scientific elements and arguments involved'; (2) Renewable Energy Partnership – 'a scheme to involve the public and private sector in the Campaign for Take-off'; (3) many others that can be easily found on the Internet and other specialized media.

As far as renewables are concerned, it is obvious that intervention is needed mainly at the local level, since the necessary legislative framework and policies are already in place. It is also of fundamental importance 'to learn the lessons of history' and use them, without forgetting the specificity of the local people and conditions. If the official policy (laws, economic instruments, and so forth) is not sufficient or effective, then, in order to make people understand better the need to use RES, a grassroots approach should be used, its goal being that of changing public opinion on this issue and of exerting influence on decisions and the policy-making process. Transdisciplinary research will have to be carried out, taking into account the country's cultural traditions and historical experience, in order to incorporate these into simple guidelines for consumers. It should lead to a more active involvement and stronger influence of the people on decisions and the policy-making process concerning the diffusion and use of renewables. A very basic question is: How can we decrease domestic energy use and costs, while at the same time contributing to protect the environment and the climate? Due to Bulgaria's lack of experience in this field, studying examples of better practices in other countries and adapting them to the specific Bulgarian conditions should be the foremost preoccupation of the present research agenda in the area of policy development for energy and environment. The priority granted to such issues by the EU/EC 6th Framework Programme for Reaserch and Development, as well as other EU programmes and initiatives, constitutes the necessary basis for a development in this direction. A couple of examples: 'ManagEnergy', an initiative of the EC Directorate General on Energy and Transport; 'Intelligent Energy for Europe', the Community's support programme for non-technological actions in the field of energy, and precisely in the field of energy efficiency and renewables.

The main conclusion is that in the case of Bulgaria, not only economic obstacles (lack of money and funding), but also a problem with priorities in the distribution of available funds, prevents renewables from being used adequately and to their full potential. There are also political and structural obstacles. Increasing the public's participation and involvement in the decision and policy making process has to be the top priority. And this is all the more relevant because it does not require big capital investments or high operational costs. Over the long term, the advantage of renewables, compared with the dwindling conventional sources, is their superiority in the field of future energy development. This aspect must be considered in detail. In addition, again in the long term, the massive use of renewables can positively influence economic development by intensifying investment, both foreign and domestic, creating new jobs, facilitating technology transfer, etc. thus contributing to the sustainable development of Bulgaria.

Note

The author is very grateful to Dipl.Eng. Valden Georgiev for his valuable help in the course of the preparation of this chapter. Thanks are also due to my colleagues at the Department of Atmosphere and Hydrosphere Composition of the National Institute of Meteorology and Hydrology, in particular Prof. Dr. Dimiter Syrakov and Dr. Stanislav Bogdanov, for their support and useful comments.

References

European Commission (1997), 'Energy for the Future: Renewable Sources of Energy – White Paper for a Community Strategy and Action Plan', COM(97)599.

Office for Official Publications of the European Communities (1994), 'The European Renewable Energy Study: Prospects for Renewable Energy in the European Community and Eastern Europe up to 2010', Main Report, Luxembourg, Office for Official Publications of the European Communities, 1994.

Office for Official Publications of the European Communities (2000), 'European research in action: Renewable energy sources – the path to the future, Office for Official Publications of the European Communities', Luxembourg, 2000.

Reiche, D. (ed.) (2002), *Handbook of Renewable Energies in the European Union I: Case Studies of all Member States and Handbook of Renewable Energies in the European Union II: Case Studies of all Accession States*. Peter Lang GmbH, Frankfurt/Main, Germany.

Websites

Bulgarian Ministry of Energy and Energy Resources, www.doe.bg.

Executive Agency on Energy Efficiency, http://seea.government.bg.

Intelligent Energy for Europe, http://europa.eu.int/comm/energy/intelligent/index_en.html.

International Climate Symbol, www.saveourclimate.org.

The 3rd Bulgarian National Communication to UNFCCC, www.unfccc.int/resource/docs/natc/bulnc3.pdf.

The 3rd National Conference on Renewable Energy Sources, www.senes.bas.bg/tretank/.

The Bulgarian National Capacity Self-Assessment Project, http://chm.moew.government.bg/ncsa/documentsEn.htm.

The European Green Electricity Network, www.greenelectricitynetwork.org/english/4_5.html.

The International Conference for Renewable Energies, www.renewables2004.de.

The 'ManagEnergy' initiative of the EC, DG Energy and Transport, www.managenergy.net.

The 'Science cafes', www.euroscience.org/MBSHIP/Bulletin/Euro25.pdf.

The Second World Renewable Energy Forum, www.world-renewable-energy-forum.org.

The UNEP Sustainable Energy Finance Initiative (SEFI), www.energy-base.org/site_fs.htm.

White Paper for a Community Strategy, http://europa.eu.int/comm/energy/res/index_en.htm.

Chapter 10

Present Situation and Future Challenges of the Estonian Energy Sector

Olev Liik

(Country Background)

Estonian power engineering has a long history and distinguished traditions. Electric lighting was first used in factories in 1882. The first industrial power plant was built at the Kunda cement factory in 1893 and the first public power plant in Pärnu in 1907. The year 1918 is regarded as the founding year of the Estonian power system. The first national electrification programme was developed in 1930. Until World War II the sources of electricity were thermal power plants that used local peat and oil shale, and numerous small hydro plants. The era of oil shale-based power production began in the 1950s and two of Estonia's oil shale power plants are still the world's largest.

The recovery of political and economic independence in 1991 brought about drastic changes in Estonia's economy. For the energy sector, these changes meant a dramatic rise of fuel and raw material prices, a decrease in energy consumption and electricity exports, but also problems with imports of oil products from Russia. A decisive factor that helped the energy system survive through the difficult first years was the fact that all necessary electricity was produced locally and 99 percent of it came from oil shale.

oil shale.
gave some potential for
Russian oil dependency

Basic Energy Data

The development of primary energy supply and final energy consumption as well as electricity and heat production and consumption, are shown in Figures 10.1, 10.2, 10.3 and 10.4.

Indigenous fuels (oil shale, wood and peat) form approximately 2/3 of the primary energy supply of Estonia. The share of renewable energy sources (mainly wood) was 10.5 percent in 2000 (see Figure 10.5). Estonian oil shale is rather unique, its reserves are the largest commercially exploited deposit in the world. Oil shale is characterized as a low-grade fuel with a low heating value (average 8.6 MJ/kg). It is a sedimentary formation, which consists of organic matter or kerogen, carbonate matter and sandy-clay minerals (18-42 percent). It contains 1.2-1.7 percent sulphur, mostly as organic and pyretic (Kallaste, Liik, 1999).

Dramatic fall in
avg supply + consumption
in 199

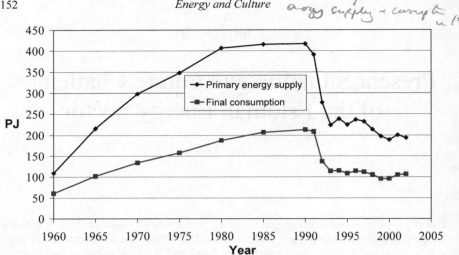

Figure 10.1   Primary energy supply and final consumption from 1960-2002
(199 PJ = 55.2 TWh)

Dramatic falls in 1991

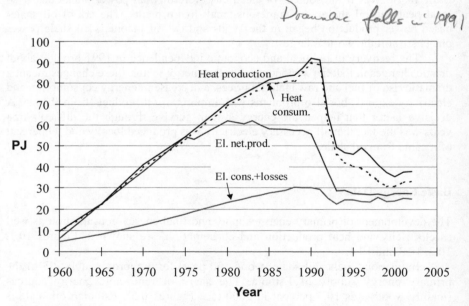

Figure 10.2   Estonian electricity and heat production and consumption from
1960-2002

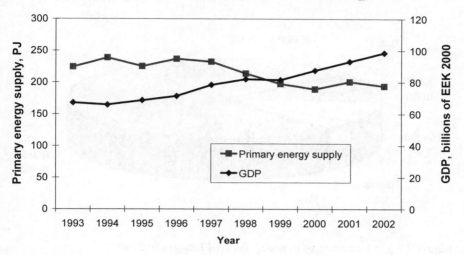

**Figure 10.3   Primary energy supply and GDP of Estonia from 1993-2002
(1 EUR = 15.64664 EEK)**

falling energy supply but rising GDP in 1990s

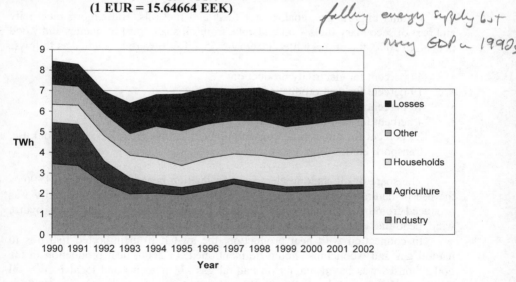

Figure 10.4   Structure of Estonian electricity consumption

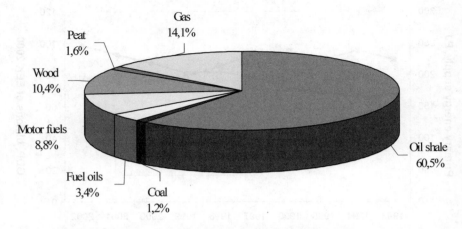

Figure 10.5   Structure of primary fuels of Estonia in 2000

Estonia imports gas, coal, motor fuels and fuel oils, and exports electricity and part of secondary fuels – oil and coke from oil shale, peat briquettes and wood pellets. In 2000, the primary fuels (199 PJ = 55.2 TWh) were consumed as follows:

*   44 percent for electricity production;
*   19 percent for heat production;
*   13 percent for production of secondary fuels;
*   4 percent for non-energy purposes;
*   20 percent for direct final consumption (industrial processes, household use, transport, etc.).

The share of oil shale in electricity production began to decrease in 1996 as the use of natural gas began to rise. In 2000 about 91 percent of electricity was produced from oil shale and circa 7 percent from gas. The other resources (hydro, wind, peat, fuel oils etc.) accounted for a total of 2 percent.

In connection with heat production, the switch from imported fuel oils to natural gas and woodchips should be mentioned. In 2000, heat production in the boiler houses was based mainly on natural gas (38 percent) and local fuels – oil shale, wood, peat and shale oil (over 40 percent).

About 12-14 percent of electricity and one third of heat is produced in the combined heat and power plants (CHP). The share of district heating in heat consumption is approximately 70 percent.

The decrease of energy intensity in Estonia's economy is an extremely positive development. GDP has increased substantially after 1994, while energy consumption has slightly declined. The energy intensity of GDP has decreased from 2.1 kWh/EEK$_{95}$ in 1993 to 1.0 kWh/EEK$_{95}$ in 2000 (Liik, 2002). Still, the level achieved is not satisfactory; and a higher efficiency in energy production, transmission, distribution and consumption are the priorities for the future.

After economic restructuring, energy consumption in industries, transport, in particular in agriculture has decreased and households are now the largest group of energy consumers (see Figures 10.6 and 10.7). Household consumption includes also private cars (see Figure 10.8).

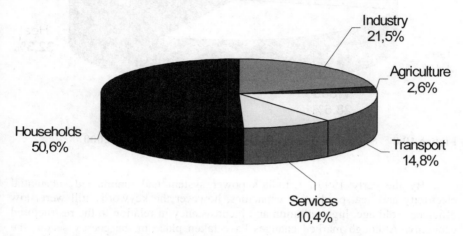

Figure 10.6   Structure of final energy consumption by groups of consumers in 2000

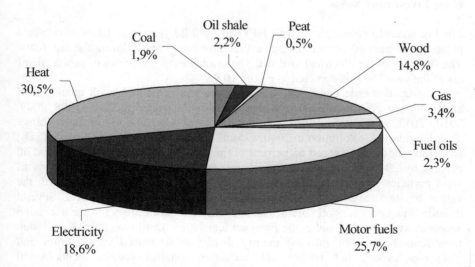

Figure 10.7   Structure of final energy consumption by energy carriers in 2000

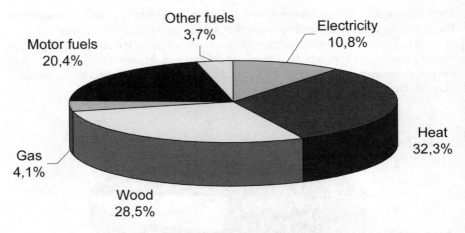

Figure 10.8   Structure of household energy consumption in 2000

Ned for renewshart in ESI

By the early 1990s, Estonia's power system had maintained substantial electricity and heat production capacities, however the keywords still were: low efficiency, old age, high pollution and inconsistency in relation to the restructured economy. Although marked changes have taken place in the energy sector, the bigger challenges are yet to come.

Huge Investment Needs

The last sizeable power plant (Iru CHP) was built 20 years ago. Units of oil shale plants are at least 40 years old; and several have been closed during the last years. The structure of the electrical grid with the available net capacities of power plants (as of the year 2003) is depicted in Figure 10.9.

Today domestic consumption is adequately covered and small quantities of electricity are exported, but severe restrictions are expected in 2005, 2008, 2009, 2010, 2012, and 2015. When signing the Genf Convention on Long-Range Transboundary Air Pollution in 2000, Estonia agreed that its sulphur dioxide (SO_2) emissions should not exceed 65 percent of the 1980 level by the year 2005, and 60 percent by 2010. The Estonian Environmental Strategy states that emissions of solid particles must be reduced by 2005 by 25 percent in comparison with the values of 1995, and NO_x emissions should not exceed the 1987 level. Several definite limits on emission volumes and specific emissions are set in the accession agreement with the EU and in the Estonian legislation. Until now the SO_2 emission constraints have been fulfilled mainly thanks to decreased consumption and electricity export, but further SO_2 reduction requires special technological measures, like change of oil shale combustion technology, fuel switching, use of renewables, etc. No problems exist with regard to fulfilling of UNFCCC Kyoto

Pollution abatement regulators + targets

not Kyoto SO₂

Protocol commitment on CO_2 reduction (8 percent decrease in 2008 as compared to 1990), since over 50 percent of emissions have already been reduced.

International

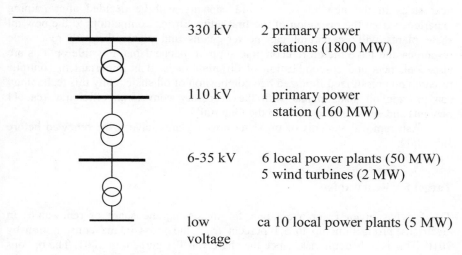

330 kV	2 primary power stations (1800 MW)	
110 kV	1 primary power station (160 MW)	
6-35 kV	6 local power plants (50 MW) 5 wind turbines (2 MW)	
low voltage	ca 10 local power plants (5 MW)	

Figure 10.9   Structure of the electrical network of Estonia in 2003

Starting from 2008 Estonian power plants have to comply with the EU directive 2001/80/EC on the limitation of emissions into the air from large combustion plants. During the accession negotiations with the EU Estonia was granted some transition periods. Existing oil shale pulverized combustion units will be phased out gradually and they can work until the end of 2005, but the total annual SO_2 from all shale power plants cannot exceed 25000 tons starting from 2012 (the actual level in 2003 was 81520 tons [6]).

After 2015 only 6 percent of the capacity of power plants that existed in the 1990s (over 3000 MW) will be in operation. This means that billions of Estonian kroons (EEK) will have to be invested over a short time in new generating equipment. It has been suggested that the total installed capacity should never decrease below 2000 MW. Considering the economic growth targets, a reduction of electricity demand cannot be envisaged.

In addition to the investments in power plants, substantial funds are required to develop and modernize the electrical networks and to implement environmental projects as well. About 1 billion EEK is required annually to reach international supply quality standards by 2012. 90-95 percent of electric grids must be renewed before 2030. It is very important to develop cooperation and interconnections with neighbouring power systems (e.g. Baltic Ring, Estonia–Finland Cable, etc.).

To fulfil the environmental requirements by the year 2005, reconstruction of two production units of the oil shale power plants with the total net capacity of 390 MW and renewal of ash filters of all units had to be completed by 2004. The new units use circulating fluidized bed combustion technology that raises conversion efficiency from 29 percent to 34 percent and minimizes sulphur emissions. The next steps in the new effort to build capacity will be decided after gaining experience from the operation of the first units. Strong competitors to the new oil shale plants will be natural gas power plants and plants that use renewable resources and have therefore economic support mechanisms. Possible options are also coal, peat and co-combustion of different fuels. It is important to continue research on pressurized fluidized bed combustion of oil shale. Only this technology can provide oil shale plants with the necessary conversion efficiency (ca. 44 percent) and emission reduction in the long run.

Ash removal systems of oil shale power plants have to be renewed before July 2009.

Target for Renewables

The EU has given Estonia a target for increasing the share of renewables in electricity production: up to 5.1 percent from the gross inland consumption by 2010. This is a difficult task, since the share was 0.1 percent in 2001. The options are: the use of biomass, wind generators and restoration of the small hydro plants that once existed. However, the hydro option is very limited because the real potential is only ca 30 MW. Hydro could contribute to a large extent if an agreement is reached with Russia on the joint operation of the 123 MW plant on the border river Narva.

In 2002, the first wind farm of 3x600 kW capacity was built in Virtsu and restoration of the present biggest hydro plant (1.1 MW) was completed in Linnamäe. In 2003, one 250 kW wind turbine was connected to the grid. It can be expected that 25 MW of wind turbines will be in operation by the first half of 2005, and an additional 140 MW will be under construction.

The renewable energy options are all quite expensive and there are serious technical limitations to wind power development (Liik, Landsberg, 2003).

The Energy Sector Has To Be Efficient

The energy sector is the basis for the rest of the economy and cannot be separated from environmental and social issues. Considering also the sector's operation costs and investment needs, one can easily conclude that energy system operation and development have to be optimal.

The first long-term national energy programme after World War II was drawn up in 1989. Since then, numerous plans at different levels have been developed. Of the most important ones, the following could be listed: 'General Principles of the Development of Estonian Power Engineering until 2030' (1990),

'Energy Master Plan for Estonia' (1992-93, international project), 'Energy Strategy for Estonia' (1996-97, EU PHARE project), 'Long-Term Development Plan for the Estonian Fuel and Energy Sector' adopted by the Parliament in 1998, the EU PHARE financed programme 'Energy Planning for Municipalities' (1998-2000) that consisted of 20 different planning projects, and 'Action Plan for the Restructuring of Estonian Oil Shale Power Engineering 2001-2006' (2001). Several international projects on environmental emissions have also been drafted, where the energy system development projections have played a key role.

In 2002, the 'Development Plan for Estonian Electrical Power Engineering until 2030' and the 'Long-Term Development Plan for the Energy Sector of the City of Tallinn (2002-2017)' were finalized. It is envisaged that the main energy policy document – 'The National Long-Term Development Plan for the Fuel and Energy Sector until 2015 (with a projection until 2030)' will be adopted by the Parliament soon and updated every three years. The national strategy 'Sustainable Estonia 21' was completed in 2003 as well.

Most of the listed documents have been developed completely or partially by the specialists of Tallinn University of Technology (TUT).

Another Capacity challenge: Changing policy states, advice + consultation.

Development of the Legislation and Regulatory Framework

Independent Estonia has undertaken the drafting of legal acts, and the creation of governmental programmes, etc, while moving from the Soviet system of standards to the international ones. Until July 2003, the most important law in the energy field was the Energy Act, adopted by the Parliament in 1997. The Electric Safety Act, the Law on Energy Efficiency of Equipment, the Law on Minimum Reserves of Liquid Fuels, etc., regulate narrower areas. In 1992, the Government adopted the Energy Conservation Programme and in 2000 this was updated as the Target Programme of Energy Conservation. The independent Energy Market Inspectorate, the main regulator in the energy field, was established in 1998.

To harmonize Estonian legislation with the EU directives and to improve the regulation of dynamic energy markets, the Energy Act was replaced by four separate laws: the Electricity Market Act, the Natural Gas Act, the District Heating Act, and the Liquid Fuels Act. The drafting of these laws took more than two years; they were finally adopted by the Parliament on February 11, 2003 and came into force on July 1, 2003. Also, the Electric Grid Code was elaborated as a supplement to the Electricity Market Act.

The Energy sector is strongly influenced also by environmental legislation, like the Sustainable Development Act, the Atmosphere Protection Act, the Pollution Fees Act, the Environmental Strategy, etc.

Estonia has ratified several international agreements, such as the European Energy Charter Treaty, the United Nations Framework Convention on Climate Change (UNFCCC) and its Kyoto Protocol, the Convention on Long-range Transboundary Air Pollution and its protocols, and the Vienna Convention for the Protection of the Ozone Layer.

Role of EU and international policy, rather than Soviet ones.

Electricity production from renewable sources is subsidized. The network companies have a purchase obligation with a feed-in tariff 1.8 times higher than the annual average selling price of large oil shale power plants. These subsidies are in force for 7 years after the construction of hydro and biomass power plants and after 12 years in the case of the other renewable technologies. All subsidies will end on 31 December 2015.

Elaboration of a comprehensive and optimal system of energy and environmental taxes and subsidies is an important task for the near future.

The contribution of TUT's specialists in the development of energy field legislation, standards, terminology, etc. has been significant.

Liberalization of the Electricity Market

Liberalization of the electricity market means opening electricity production and sales to competition, while transmission and distribution remain natural monopolies. The Estonian electricity market has been open since 1999 to eligible customers whose annual consumption exceeds 40 GWh. These consumers have a right to purchase electricity from any producer or seller in the market and an obligation to pay for network services. Consumption by eligible customers currently accounts for circa 10 percent of the total consumption. During the accession negotiations, Estonia and the EU reached a compromise solution for further step-by-step opening of the electricity market. At least 35 percent of the market must be opened before December 31, 2008, and for all non-household consumers (ca. 77 percent) before December 31, 2012. The market will operate according to the rules of the new Electricity Market Act and the Grid Code.

It is a widely accepted opinion that liberalization will increase the efficiency of the system and the quality of its services. Reductions in consumer prices are probably only short-term. Still, an open market also creates new problems. In Estonia's case, the main risks are:

- The market will not be really open, if the number of independent producers and sellers is too small (i.e., offering no competition for Eesti Energia Ltd.);
- A shortage of generation capacity can occur and prices will rise, if the market participants (also in neighbouring countries) only want to sell and buy, but not invest – this is the most serious risk;
- Estonia is so small that large-scale cheap imports can destroy local production and investments, and make the country dependent on neighbouring states;
- New power plant investments can increase the share of imported fuels (natural gas technologies are cheaper, more efficient and environment-friendly than the other fossil technologies), cause supply security and price risks, and worsen the foreign trade balance;
- Low interest of large-scale customers in small producers and the cheap electricity imports can slow down the development of cogeneration;
- The pressure to raise the electricity prices of the closed part of the market (especially households) will increase;

• Considering the small size of the Estonian electricity market, the complexity of power system control, the costs of operating the market, the volatile prices and the possible decrease of supply security and reliability due to insufficient investments in the whole region can easily obliterate the expected positive effect of liberalization. *Small market size*

Opening the electricity market also causes institutional changes in the energy companies: production, network and sales activities have to be separated from each other. The national grid and existing strategic power plants and oil shale mines will remain under state ownership. *benefits of power liberalist*

It has to be mentioned that between 1995 and 2001, all activities of the power sector were influenced by the negotiations involved in selling 49 percent of the shares of the two biggest oil shale power plants to US capital (NRGenerating International B.V.). This deal met strong opposition among energy specialists and also among the general public. In spite of the political decision to sell the shares, the privatisation process failed at the end of 2001. *international capital flows in tension with a bulky investment a system*

Forecast of Main Energy Indicators

The development of main energy indicators until 2010, as forecasted in the 'Draft National Long-term Development Plan', can be found in the following table. The tendency will be towards a stable primary energy supply, more efficient energy conservation and use, increase in the share of renewables and CHP, increase of electricity consumption, liberalization of the electricity market and the reduction of emissions.

Table 10.1   Estonian main energy indicators until 2010   *Trends in 2005*

	2000	2010
Primary energy supply (PJ)	189	220-250
Consumption of oil shale (Mt)	13.2	11-13
Share of renewables in primary energy supply (%)	10.5	13-15
Share of renewables in electricity generation (%)	0.1	5.1
Final consumption of electricity (TWh)	5.4	6.5-8.0
Necessary net capacity of power plants (MW)	1980	2200-2500
Share of CHP in electricity generation (%)	12-14	15-20
Maximum net load of Estonian power system (MW)	1400	1600-1900
Openness of electricity market (%)	10	35-40
Heat consumption (TWh)	8.5	8-9
Share of CHP in heat production (%)	33	35-40
SO_2 emissions (% from limit in 2008)	181	90-100
CO_2 emissions (% from limit in 2008)	48	50-55

Conclusions

Estonia's energy sector and especially power engineering face serious challenges during the coming years:

- To secure a reliable, high quality and affordable energy supply in a situation in which the majority of existing production capacity and energy networks have to be closed down or restructured;
- To fulfil environmental restrictions and obligations;
- To fulfil the targets for increasing the share of renewable energy sources;
- To develop the cogeneration of electricity and heat;
- To open the electricity market;
- To fulfil the national reserve requirements of liquid fields;
- To increase the efficiency of energy conservation, transmission, distribution and consumption;
- To construct the interconnections with the Nordic Central European power systems;
- To participate in the development of new technologies and fuels;
- To elaborate optimal system of taxes and subsidies;
- To enter the GHG emissions trading market;
- To maintain the know-how and to develop education and scientific research in the energy fields.

    It will be interesting indeed to see how Estonia and its energy sector meet these challenges.

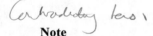

Note

All statistical data in this chapter has been provided by the Statistical Office of Estonia.

References

Draft National Long-term Development Plan for the Fuel and Energy Sector until 2015 (with a projection until 2030). Report to Ministry of Economic Affairs and Communications of Estonia. Tallinn: Tallinn University of Technology. (In Estonian.)
Energy Balance (2000), Tallinn: Statistical Office of Estonia, 2001.
Energy Balance (2002), Tallinn: Statistical Office of Estonia. 2003.
Kallaste, T., Liik, O. and Ots, A. (eds.) (1999), *Possible Energy Sector Trends in Estonia. Context of Climate Change.* Tallin: Stockholm Environment Institute Tallinn Centre, Tallinn Technical University: www.seit.ee/download/possible_energy_sector.zip.
Liik, O., Landsberg, M. and Oidram, R. (2003), 'About Possibilities to Integrate Wind Farms into Estonian Power System', *Proceedings of the Fourth International Workshop on Large-Scale Integration of Wind Power and Transmission Networks for Offshore*

Wind Farms, 20-21 October 2003, in Billund, Denmark. Royal Institute of Technology KTH, Stockholm.

Possibilities for increasing the share of renewables in the electricity production in Estonia. Report to Ministry of Economic Affairs and Communications of Estonia. Department of Electrical Power Engineering of Tallinn University of Technology, Tallinn.

Chapter 11

Energy Efficiency and Lifestyle

András Zöld

(handwritten: Hungary / ?)

The well-known problem of environmental pollution and limited fossil energy sources, on the one hand, and energy consumption, on the other hand, has prompted the EC to issue the Energy Performance Directive 2002/91/EC (EPD). Each Member State of the EU25 is obliged to develop and implement the national regulation by January 4, 2006. Regulations should follow the concept of the EPD and they should harmonize as much as possible, although different climatic and social conditions will inevitably be reflected in input data as well as in the form and numeric values of the requirements.

All legislative measures aimed at the promotion of sustainable building are appreciated by responsible professionals and decision-makers. At the same time, it is their responsibility to detect possible misinterpretations or inadequate drafting of the requirements.

With regard to the EPD, two main problems should be mentioned. On the one hand, although in some sentences a comment on the rationality of energy saving investments can be found, the EPD deals mainly with operational energy, thus the life-cycle energy balance is neglected. On the other hand, the complex means of expressing integrated operational thermal performance include many components which are not necessarily building-related. The term 'integrated' means that all kinds of energy consumption are included: besides the 'classical' uses, such as energy consumption employed for heating, ventilation and cooling, it includes also the energy consumption for hot water, lighting and households appliances. *(handwritten margin note: life cycle neglected)*

(handwritten note: Regulation creeping into more areas)

Without a doubt, the intention to cover and regulate all kinds of energy consumption in buildings is a considerable step forward. However, it raises several crucial questions, above all in residential buildings where the relative weight of the user-related components seems to exceed that of the building-related ones. This paper focuses on this problem.

Each component of energy consumption is to be expressed in primary energy. This is obviously the correct approach. Nevertheless, the 'exchange rate' of the primary energy increases the relative weight of the user-related consumption and makes the comparability uncertain if different numerical values are used in different regions.

Users' Role in Building Regulation

The more numerous are the factors taken into account, the more complex the design and the sounder the end results will be. However, it is questionable whether the necessary input data are available and whether cooperation between architects, engineers and future occupants is possible in the very first stages of the design.

Architects and engineers employ tools to limit heat losses, to increase utilizable solar gains, and to improve the efficiency of service systems. With regard to energy consumption for heating, ventilation and cooling, there are conventional design values, such as a set indoor temperature and air change rate. Although real consumption depends on users' behaviour, the expected (or 'design') consumption can be calculated and the performance of different buildings can be compared on the basis of the same conventional or standard design input values on one hand, and, on the other hand, specific output values, such as consumption per annum and floor area (or the like).

The energy consumption of household appliances is more uncertain; variations in their type and power cannot be predicted, nor can changes in the use of the building itself. Even the specific number of occupants per square metre and the time they spend at home show wide variations, not to mention the type of required power, the schedule of use of different household appliances, and the consumption of hot water. In this sense, there is no conventional or normative user. Comparable values can hardly be expected due to the social differences in different regions.

Defining the 'conventional user' is not simple. Should different social situations be reflected in the national regulation? Should the comparability of energy performance of houses in different regions be abandoned? Would stricter requirements have a stimulating effect on improving households and raising awareness of an energy- and environment-conscious lifestyle, or would unrealistically strict requirements only result in disappointment? Does an energy-conscious lifestyle mean that one's comfort or hygienic standards must be lowered? Or does public awareness of energy issues open promising new perspectives in the long run?

Simplified Analysis of the Energy Balance of a Building

In a given climatic region the basic degree-hours value (Figure 11.1) is:

$$DH_{or} = \int_{t_e \le t_i} (t_i - t_e(\tau))d\tau$$

The yearly function of the external-temperature, t_e, characterizes the climatic region. The indoor temperature t_i is a user-related value: for residential buildings 20°C is the conventional figure. The degree-hours depend on the set indoor temperature; in a temperate climatic region a $\pm$ 1 K modification in t_i results in $\pm$ 8 percent change in the degree-hours. The milder the climate, the greater the effect of the set indoor temperature. The length of the heating season when the auxiliary heating system must be operated and the degree-hours for the heating season depend on the balance point.

A modest 1-2 K change in the balance point temperature results in a change of 25-45 days in the length of the heating season.

Traditional energy-saving methods, such as setting back the thermostat at night, partial heating (in space) and the metering of heat consumption in flats of residential buildings, are less efficient in low-energy buildings: the estimated saving of intermittent heating is 2-4 percent. With the use of good thermal insulation, the time constant is high, and the cooling down is slow. Again, as a result of good thermal insulation, the thermal coupling of neighbouring flats is higher than that of the flat and the environment. If emitters in some of the premises are switched off, their indoor temperature decreases only by a few Kelvin degrees: heat flow from the neighbouring rooms will be comparable with the heat flow towards the environment. If individual heat metering per flat is applied, the measured value may be precise, but there is no way of knowing how the consumed heating energy is shared between the metered flats and the neighbouring ones.

A building is exposed to solar radiation; the utilized solar gain covers part of the heat losses. The resulting balance temperature of the *empty* building depends exclusively on building-related characteristics, such as: the basic heat loss coefficient (taking into account only spontaneous air infiltration), orientation, solar access, energy collecting elements of the building, and heat storage capacity.

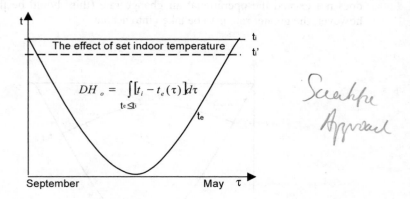

$$DH_o = \int\limits_{t_e \leq t_i} [t_i - t_e(\tau)]d\tau$$

Figure 11.1   Basic degree-hours showing the effect of set indoor temperature

Depending on the form factor (or surface to volume ratio) and the average U value, the realistic range of the basic heat loss coefficient HLC_b is between 0.18 – 1.8 W/Km^2_{floor}.

The range of solar gains can be estimated, taking 0.10-0.14 as a typical window/floor area ratio, g as a value of 0.5, and an average utilization factor of 0.6.

In the selected climatic region (temperate zone) for a north facing or shadowed window 40 W/m^2, for a south window 75 W/m^2, average flux can be taken in to account. Dividing the utilized part of this gain by the basic heat loss coefficient, the first balance temperature difference (for the empty building) can be calculated (Figure 11.2).

In a very pessimistic case this balance temperature difference is about 2 K (high surface to volume ratio, poor thermal insulation, north facing or shadowed windows). 4-6 K is a modest estimation, nevertheless good thermal insulation together with passive solar measures may achieve even 9-11 K.

With this balance temperature difference (which changes month by month together with irradiation) a reduced degree-hours value can be calculated for the empty building. In a given climatic region, the conventional indoor temperature value depends exclusively on the thermal properties of the building and provides a comparable figure (either absolute values or the ratio of the degree-hours of the *empty* building and the basic degree-hours can be used).

The next step is to calculate the balance temperature difference for the *occupied* building. By *occupied* building we mean one containing people but no household appliances and artificial lighting.

This kind of occupation has two consequences. On the one hand, due to the metabolic heat output, internal gain must be taken into account; on the other hand, the air change rate (operational ACH) should be adjusted to the number and activity of the people present. It is supposed that the spontaneous filtration rate does not exceed the operational air change rate (this should be the normal case, however, the greater rate is to be taken into account).

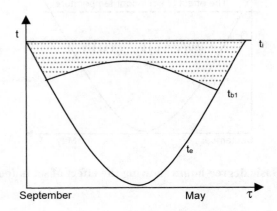

Figure 11.2   The case of an empty building: the effect of solar gain

Regarding ventilation losses, it is an arguable question whether these depend on the design or not. The necessary air change must be provided and this depends on the number and the activity of the occupants. No doubt, ventilation losses depend on the building characteristics. If the spontaneous air change by filtration exceeds the necessary air change rate; if in order to prevent fabric damage and mould growth (as a consequence of poor insulation and thermal bridges) the air change rate must be kept higher; or if the contaminant emission of building materials requires a higher air change.

We now examine the operational air change rate. The remaining question is, how will fresh air be heated up.? When calculating the operational heat loss coefficient, HLC_{op}, the heat recovery system should be taken into account.

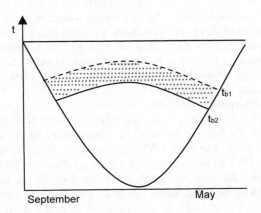

Figure 11.3   The effect of the metabolic heat gain

The balance for the occupied building depends first of all on the number and activity of occupants and may exhibit a very wide range of numerical values. Considering 1 person for 20 m² floor area, present 100 hours a week, with a modest metabolic rate of 100 W and a ventilation heat loss coefficient of 0.53 W/Km^2_{floor} (ACH = 1 with heat recovery or ACH = 0.5 without it), the estimated balance temperature difference is 3 K, although in a 'super-insulated' building this difference can be as high as 8 K. Clearly, the specific number of occupants per floor area and the time they spend in the building may vary: a conventional value might be half, as well as double the data above, which means an average specific output of 3 W/m^2. Thus the simple presence of people in the building has a considerable effect on the remaining degree-hours value, depending on their numbers and the time spent inside. The better the thermal properties of the building, the greater this effect.

Use p has big effect

The balance temperature difference, due to the metabolic heat of occupants may be of the same order as that due to solar radiation. There is no convention to express the metabolic heat flow in terms of primary energy.

Considering the balance temperature differences for the empty and occupied building a degree-hours value can be determined. The sum of heat losses, calculated with HLC_{op} operational heat loss coefficient for the remaining degree-hours, is covered by the internal heat gains (household devices, lighting) and the heating system.

$$DH_{b2} = \int_{t_e \le t_{b2}} (t_{b2}(\tau) - t_e(\tau))d\tau$$

There are several sources of heat in an inhabited building. Among them only artificial lighting could be influenced by the design, by appropriate use of day lighting, i.e. by carefully choosing the combination of room depths, position and size of windows and roof lights – the rest is not connected to the building itself.

In the household there are appliances, the total energy consumption of which is considered 'space heating flux' (iron, refrigerator, etc.), while other items have a limited space heating effect (e.g. hot water consumption – except in the rare case when heat recovery is provided from the waste water). Here the 'net heat flow' should be accounted as heating power.

Certainly internal gains from the household exist all year round, but they have some positive effect only if $t_e < t_{b2}$, otherwise they are superfluous from the point of view of the heating space. The contribution of these gains to heating comes at a high price. For each kWh 'heating flux' from lighting and household devices multiple kWh of primary energy consumption are to be taken into account, and for the rest of the year, the energy consumption of the household is on the negative side. Moreover, the same physical phenomenon which provides some contribution to heating in winter is very negative from the point of view of summer overheating.

Having, on average, 5 W/m^2 consumption of lighting and household devices, the total heating contribution per year is approx. $150 \times 24 \times 5/1000 = 18$ kWh/m^2year, the 'price' in primary energy consumption being $365 \times 24 \times 5 \times 3/1000 = 131{,}4$ kWh/m^2year. Here it was supposed that the length of the heating season, determined by t_{b2}, is 150 days; in a well-insulated building this period would be shorter.

The better the thermal properties of a building, the greater the effect of the internal gains from the household devices. Just as in the number of occupants, there may be wide variations in the number, the type and the use of household appliances and lighting. Depending on input data or conventional figures the related balance temperature difference can be between 1K and 5-6 K. In the integrated energy consumption the 'exchange rate' of primary energy is another factor of high importance beyond the scope of architects and design engineers.

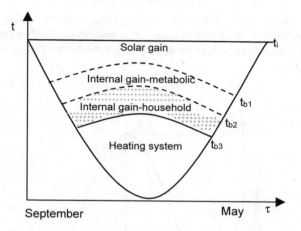

Figure 11.4   The effect of internal gains – household

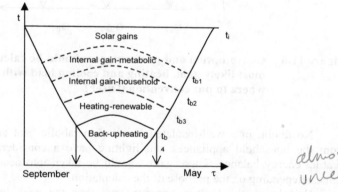

Figure 11.5   Bivalent heating and the length of the heating season

Having considered the third balance temperature for the 'inhabited' building (Figure 11.3), the rest of the degree-hours are to be covered by the heating system. For that purpose mainly primary energy is used (mostly natural gas, sometimes oil, or coal). Electric heating is penalized by the 'exchange rate' of primary energy. For heat pumps that rate and the heat loss coefficient approximately compensate each other. Solar gains (other than direct gain or for the preheating of fresh air – these have already been taken into account) or geothermal sources can reduce the primary energy consumption. Biomass or wood may be used as renewable fuels. In any case the relatively small amount of the consumption of pumps and fans is to be considered unless simple stoves or radiator heating is used. If this is the case, one more balance temperature difference can be defined, taking into account the use of

Energy and Culture

a heating system with renewable energy (e.g. active solar thermal with instantaneous output or stored energy, or a heat pump with evaporators in the outdoor air). The last part is to be covered by a back-up system, using fossil fuel (Figure 11.5).

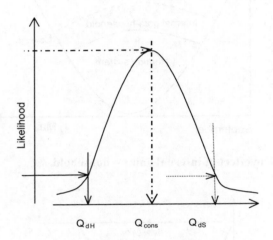

Figure 11.6   Consumption and performance should be calculated with the most likely gain, heating and cooling load with acceptable risk – where to put conventional user?

No doubt, in a well-insulated building metabolic heat and the heat output from the household appliances and lighting have a considerable effect on the heating energy balance. These gains should be taken into account with different values, depending on the purpose of the calculation.

Calculating the built-in capacity, and thus the peak load of the heating system, we must be on the safe side: a modest heating contribution from the internal gains is to be included. The input data should correspond to an acceptable risk level, e.g. 10 percent (the likelihood of a lower value is less than 10 percent). Overestimating this contribution leads to underestimating the peak load. The point is to avoid the problem of an undersized heating system.

However, calculating the annual energy consumption on the basis of these figures (low gain and corresponding higher output of the heating system) the result would most likely be false. Annual energy consumption and the integrated performance of the building should be calculated using the mean value of gains (Figure 11.6). But average internal gain will mean not only higher heating contribution but also much higher consumption in terms of primary energy (the decrease of the consumption of a heating system is typically primary energy, the contribution from internal gains is to be multiplied by the 'exchange rate' 2.5-3.0).

The same concept is to be applied for summer: calculating the risk of overheating or the built-in capacity of the cooling system, the input data of gains should correspond to an acceptable risk level, e.g. 10 percent on the opposite side.

Domestic hot water consumption should be mentioned separately: it seems to be the most important component, and at the same time the most difficult to grasp, for several reasons. It is likely the to be the largest item in the final balance. The basic values, i.e. the volume and the temperature of hot water, can hardly be influenced by the designers. The modest effect of water-saving taps and equipment can be mentioned, but it is of secondary importance compared with the number and habits of the occupants, as far as design input data are concerned. Hot water consumption has a limited space heating output (except in the rare case of heat recovery from waste water). The hot water system can be based on direct use of primary energy or on electric energy and can be supported by a solar system, and it may be the most indefinable component of the integrated energy balance, depending practically on anything but the building itself. The basic value (MWh/a), calculated from the volume and temperature difference, can be reduced by 50-70 percent: in a temperate zone this is a realistic solar contribution. The rest − or the total, if no solar system is applied − can be multiplied by the 'exchange rate' provided the auxiliary system is based on electric energy.

Whether a PV array can be applied depends on the orientation, building form, aesthetic appearance and constructional details. The extent of the coverable surface, compared to the floor area must also be taken into account. Producing electricity from renewable PV represents a great chance to achieve an excellent integrated characteristic.

The Building within the Integrated Balance

Let us suppose that according to the EPD, all of the above listed components are included in some specific value, expressing the integrated energy performance of a building in primary energy. There are − theoretically − two óptions:

Option One

There is one and *only one* specific value, expressing the integrated energy performance. Provided that the building meets the requirements of the recent national standards, that specific value depends:

- First of all, on hot water consumption, on the energy carrier used for water heating and on the household appliances and lighting; thus it depends mainly on the number of users and their habits, which cannot be defined precisely for the long term at the design stage, since they can change several times during the physical lifetime of the building;
- On the 'exchange rate' of primary energy, thus on a convention, which is as stable as a bureaucratic or political decision used to be;
- Less importantly on factors such as heating and ventilation systems;

- The smallest component is the thermal performance of the building itself. The building has the longest physical lifetime; it is the most stable element in the whole context, and it is designed by architects and engineers. At first sight one may have the impression that the building itself is negligible as its quality can be camouflaged in several ways.

Having one and *only one* integrated value with an upper limit for a given building, the requirements, based on a sum of several components, can be met with various combinations of these components. Due to trade-off possibilities, a better building shell allows a less efficient hot water system, and *vice versa*. Since building-related components of integrated energy performance are less important than other components, the thermal quality of the building (thermal insulation, air tightness, insulation, etc.) may be poor, provided that this fact is compensated by an efficient hot water system or good household appliances (the use of which depends on users' variable and unpredictable habits). The other – less irrational – way of compensating poor building thermal quality is to cover the building with PV: the higher the exchange rate of the primary energy the more deficiencies of the building shell will be covered up in both senses of the term.

Option Two

Due to the above problems it seems inevitable to develop or have at hand a parallel requirement system based on building-related parameters. The thermal performance of buildings should be expressed in the form of a specific value or values meeting the following requirements:

- Its meaning is unambiguous;
- It can be calculated from the first stages of the design;
- Is complex but simple, including all, but only building related, important components of the energy balance.

Even this requirement system is not free of variability due to users' influence; nevertheless, the previously mentioned conventional values facilitate the calculation of comparable and characteristic specific values. Regardless of any changes in the use of the building, in the number of its occupants, etc., the fulfilment of that minimum requirement system should guarantee an acceptable energy performance. The architect can only be held responsible for this aspect.

The way the heating, cooling and lighting systems harmonize with the building, and their structural and functional integration with it, including the active solar thermal and PV systems, depend on the cooperation between architects and engineers.

The result can be expressed as some 'interim integrated value', which – using a hypothetical user as input – characterizes the energy consumption of a building for each kind of energy source and includes several details, such as fans, pumps, thermal insulation of a distribution network, boiler efficiency, etc. The architects and engineers who design the building share the responsibility for creating this harmony.

To oversimplify we could say that we are dealing with two issues: the heat losses of the building, and the primary energy consumption necessary to make up for them. The ratio of these two factors is a good indication of the performance of the building and the related systems.

No doubt, the efficiency of the hot water system is the responsibility of the designing engineers. By this we mean the quantity and source of energy used for the unit volume of hot water at a given temperature. However, the engineers are not in control of hot water consumption. To limit it they would have to undersize the system, which is of course only a hypothetical solution.

The energy consumption of household appliances and office equipment is far beyond the responsibility of architects and designing engineers of the building. The responsibility of the designing engineers is restricted to the correct interpretation of the user-related data and the provision of energy supply.

First, the heat output of the household appliances and the metabolic heat of the users themselves are to be included in the calculation of the heating and cooling load as well as annual energy consumption. Here two extreme values are referred to, and loads should be calculated on the basis of the extreme design values of internal gains at an acceptable risk level. The consumption calculation should be based on the mean value.

Secondly, energy must be provided up to a rational limit for the consumers of the household. The engineers are responsible for the distribution networks, and working together with the architects they can provide an *in situ* source of renewable energy (e.g. a PV array). Nevertheless, the number of appliances, the amount of power required to run them, their annual consumption, etc., cannot be calculated in the planning phase.

Conclusions

If we focus only on operational energy consumption, we can draw these conclusions.

It is difficult to exaggerate the importance of improving the energy performance of buildings. The more articulate approach, expressed as 'integrated energy performance', seems to be promising; nevertheless, the possibilities and constraints in the design process must be taken into account, and the responsibility of the planners should be clearly stated.

By 'integrated approach' we mean that the energy balance involves several user-related factors. There is an urgent need to define conventional users, their specific number per floor area, and the conventional household. The design value of hot water consumption must also be defined.

The input data of a conventional user is indispensable from two perspectives. First, his/her presence and use of household appliances represent a contribution to heating and a component of the cooling load. These data are needed for the sizing of the HVAC systems – any discrepancy leads to a false built-in capacity of the mechanical system. Secondly, household appliances consume valuable energy – much more than their contribution to heating. Even in the cold period, their contribution to heating is penalized by the 'exchange rate' of primary energy. Consumption continues throughout the year and in the hot months heat output should be transferred at the cost of further energy consumption.

The above facts mean that the data referring to the conventional user require a statistical approach. Different data should be used for the sizing of different HVAC systems. Also, different data should be used for the calculation of energy consumption of HVAC systems and households from those used to calculate 'integrated performance'.

The 'exchange rate' of primary energy should be agreed upon, otherwise the integrated performance of buildings in different countries will not be comparable. Especially user-related consumption depends on this value, which can fundamentally distort the integrated performance. Otherwise the same buildings in the same climatic conditions and with the same users' pattern may have different qualifications, depending on their 'legal status', that is, whether they are 500 m. north or south of a border.

Among the several components of the integrated approach, the building itself must not be forgotten. Compensating for a poor building shell with better household appliances or, hypothetically, reduced hot water consumption, cannot be allowed beyond a certain limit.

It is indispensable to develop or to keep a minimum requirement system focusing only on building-related data, which can be used at the very first stage of the design and guarantees a good thermal performance of the building itself, as well as an acceptable 'integrated performance', whatever the users' habits might be.

An interim level can be considered, which includes the building, the HVAC and lighting systems.

The integrated performance can be calculated using the data referring to the conventional user; nevertheless, the responsibility for the fulfilment of such a requirement should be shared between the architect, the engineer, the user, the people in charge of housekeeping and the management.

It is perhaps the intention of the EPD to involve these actors in the realization and maintenance of a sustainable energy-conscious environment. Other portions of the EPD, such as the certification of buildings and the regular inspection of boilers, heating and air conditioning systems suggest this intention. This is an undoubtedly an important step forward: even the best buildings and the most efficient mechanical systems can be operated in an inappropriate manner. Nevertheless, all the actors should share tasks and responsibilities, which should be clearly formulated, and the regulations should be adjusted to the possibilities and constraints of the design process.

no biblog / refs !   N. Unclear to a wider audience

PART 5:
ENERGY AND RISK

Chapter 12

Social Uncertainty and Global Risks

Michalis Lianos

[handwritten: Rush / UK]

The environmental perspective seems hardly a matter of controversy. The very fact that all institutional actors, from political parties to medium-sized enterprises, tend to integrate such a dimension into their discourses and practices shows the great diffusion of a silent, unexplored consensus concerning 'the environment'. The 'stewardship' of the natural environment is a matter of such widespread consensus that it would be surprising for any institutional approach not to include an environmental aspect in its discourse. This consensus is more a common sense attitude, a tautological reaction to a prevailing trend, rather than an analytical attempt at locating and understanding the functions and hidden disparities involved in the Western interest for environmental care. *[handwritten: tautologous]*

However, it seems difficult to address these functions without dealing with a fundamental problem, namely that, although convergence on the preservation of the natural environment can mean different things and serve different purposes, it has a structural role as a unifying cultural axis. We need to think critically about this phenomenon. 'The environment' has rapidly reached a mature status as an indispensable social, cultural and political lever, whose operation crucially influences the generation and handling of many forms of power in capitalist democracies. In the last thirty years the great political and economic forces of the planet have been subjected to the critical influence of a category, which somehow stands outside their sphere of influence. *[handwritten: environment as route to power]* *[handwritten: rise / can't + decline of religion]*

It may be that the decline of religion leaves new territory for ethical considerations to be redeployed in order to challenge administrative and market institutions. Background data can be used to explore this hypothesis, particularly since there is certainly a link between different domains of belief (see Weaver, 2000). It also appears likely that this new ethical force could no longer be of a metaphysical nature and needed a tangible reflection in the outer world; but it is certainly not a coincidence that this is not the world of humans with their inequalities, but one that can be easily represented as equally beneficial to all people, yet remains voiceless and open to interpretation. The 'environment' is a world of purity and stability threatened by pollution and risk, the sum of all life that should be sustained, or an asset to be optimally managed; to different parts of society, it can be an almost mystical entity, a source of health and beauty, or a capital of high order. And to those who are farther away from power it is often a remote consideration, since they would prefer to benefit from it by exploiting and damaging it a little before they start protecting it (also see Kemmelmeier et al.,

[handwritten: political philosophy]

2002; Elliott et al., 1995) The 'environment' is an excellent tool to challenge power in an era where individual competition and personal concerns have dissipated all the momentum of collective movements.

What kind of challenge does the environmental perspective represent for power in late capitalism? This is important to know, not only in order to comprehend the conceptual and discursive struggle that determines the categories of social life, but also to understand which direction established forces prefer to take when change becomes necessary. In that sense, 'the environment' – like any successful critical discourse – is a compromise between radical critiques that meet with resistance before they reach the public sphere and unadulterated liberalism oriented towards ever-increasing competition (see Milbrath et al., 1994). The analytical potential and the meaning of this compromise will become clearer as we take a closer view at the nature of the challenge that the environmental perspective represents.

Problematics

It is painfully obvious that any use of the term 'environment' which does not apply to the study of specific ecosystems is extremely imprecise. This does not automatically mean that such a vague use has no meaning or utility in the context of social and political sciences. It is quite easy to defend that utility; it responds to the proven need for the existence of an umbrella concept, which brings under one reference the relationship between the functioning of human societies and their environment. Such a concept is indispensable in politics and in social action, as well as in theoretical debates. My point of departure here is not simply to suggest that it is necessary to address the issue of environmental thinking in an all-embracing fashion. It also seems, I think, legitimate to put forward that identifying a complicated context with a single reference is really a precondition for bringing it into public existence.

In that sense, the environmental perspective as such is a successful collective representation of a cultural 'horizon', which can supposedly seem specific enough to differentiate it from other 'non-environmental' perspectives (see Smith, 1999). By 'perspective' I mean the individual representation of a 'cultural horizon'. The distinction is essentially one between a collective thematic culture (a 'horizon') and the assimilation and use of this culture by each single socialised individual (a 'perspective'); the environmental perspective can be understood as the sum of pro-environmental views held by a socialised individual at a specific point in time. That is obviously not the same as to say that 'environmentalism' is in its inner space undifferentiated. On the contrary, there is a sustained line of literature (e.g. Wynne, 1987; Lascoumes, 1994; Martell, 1994) illustrating in detail that the preservation of the environment, like every other attempt to determine the content of the future, is a combat field for social forces, opposed interests and antagonistic perceptions related to social divisions. On the other hand, the distinction between the environmental register and other registers suggests that despite those internal antagonisms, there is a need to identify the specific cultural parameters which constitute the environmental as a distinctive field of debate and action, whether in terms of consensus or antagonism.

We should take the widest possible angle in addressing this question – not only because we are unaware of the probable extent of the questions involved – but also because issues incorporating perception and action into an indivisible cultural totality are extremely difficult to approach. For that we need the lateral, transversal axes which apply to the social in an overarching fashion, from the individual to the institutional, from self-identity to long-term cultural development, from day-to-day interaction to comprehensive political organization. In order to cover such a spectrum of formal methodological preconditions and define some fundamental questions, I will deal here with an initial comment on the restructuring of power that the environmental perspective entails for contemporary capitalist societies.

implications of envil perspective

The 'Environmental' Perception of Power Management

pompous

There is a fundamental, albeit somewhat simplistic, distinction to draw between what can be called a 'sociocentric' model of perceiving, managing and distributing power and an 'environmental' alternative. The quintessence of this *differentia* lies, I think, in the legitimating basis of the two cultural horizons involved. The sociocentric refers to society as such in order to structure the terms of defining and managing social relations. The environmental establishes a social externality by looking at human society as a subsystem that should ultimately comply with the priorities of the natural environment to which it belongs. The ethical dimension of the latter world view is undoubtedly of great importance, but it would be best understood only as a prescriptive basis of reasoning. The consequences of this reasoning are in social terms more significant than the origins. *human society must fit in*

The existence and development of a particular cultural horizon is by definition the outcome of social relations built against a background of biological assimilation, what Sperber calls 'an epidemiology of representations' (Sperber, 1996). In that sense, the mere existence of the environmental horizon constitutes *?!* sufficient proof of an equilibrium of forces in a given society that cannot ignore the utopia of adapting social organization to a vision of natural harmony. Allan Michaud (1989) supplies an analysis of the environmental horizon as a carrier of *Unintelligible* cultural change. *cult as agent of change*

Establishing the externality of the natural environment has massive *historical* consequences in regard to power. First and foremost, *it is a historical novelty that novelty* *the organization of human societies is now subjected to another priority than that* of the human species or some metaphysical entity. Although it is true in several cases that the divine has represented for some cultures a link with the natural *cf.* environment, it should not be overlooked that some mediation from an *religion/* anthropocentric point of view is always involved, since access to the divine is *belief* obtained through belief. In a rather reverse manner the 'environmental' perspective develops heteroreferentiality as a critique to anthropocentrism and sociocentrism, and in doing so builds a radically new grid for understanding, regulating and modifying social relationships. See for example Castree's argument (2002) that

critique of anthropocentrism

both policy-oriented and 'critical' environmentalism are still too anthropocentric. A deeper 'ecocentrism' is proposed as a necessary step towards a more sincere commitment to the environment.

At the same time, an opposite thesis is often put forward. Beck's position is representative of a whole line of arguments that focus on integrating the natural and the social in a unified undifferentiated context. According to him, it is already becoming recognizable that nature, the great constant of the industrial epoch, is losing its pre-ordained character, is becoming a product, the integral, shapable 'inner nature' of (in this sense) post-industrial society. The abstraction of nature leads into industrial society. 'Nature' becomes a social project, a utopia that is to be reconstructed, shaped and transformed. *Renaturalization* means *de*naturalization (Beck, 1994).

Although it is difficult to disagree with his conclusion, the whole account of the relationship of the dyad natural/social in a rather perfect correspondence to the dyad industrial/post-industrial seems quite schematic. The production of representations that are sensitive to nature and to the ways in which nature should be handled, shows primarily an urgency to construct limits between the social and the natural in order to regulate, correct and tame the former according to the needs of the latter. Hence the epistemological and sociological discourse that argues for a fusion of the analytics of the natural and the social environment. (For an interesting critique on the depoliticizing dimensions of that discourse, see Caillé, 2001; for an argument on the unifying social, political and economic characteristics of environmental issues, see Saurin, 2001) Asserting that 'renaturalization is denaturalization' is then only true in a context where 'natural' means unprogrammed. *In that sense, the occurrence of a nuclear accident is far more natural than protection in a reserve of a species facing extinction.* However, in the elaborate framework of contemporary ideology and sociality it would be too simplistic to accept a definition of the 'natural' that takes so little account of the social and political issues involved in today's conflict over representations. It is not enough to focus on the fact that nature has come under human control and is therefore 'socialised'. There are major social implications connected to that shift, which have to do with the development of a whole cultural device designed to confront and organize the 'responsibility', as Jonas defined it (1984), resulting from human domination over nature.

Both necessary and interesting is the task of looking into the modalities of this new intentional construction of a space, i.e. the space of representations for nature that are meant to address both ethical and technical aspects of the subject. Identifying, i.e. defining, nature in the post-industrial society, is not a neutral theoretical quest. It is an obligation that the environmental cultural horizon has imposed and rendered politically meaningful; for, to consider 'nature' is to respect and preserve it, or at least to pretend to do so, even for those who would prefer to have no constraints in their activities, may they be inveterate car users or transnational chemical companies. In regard to the natural environment, to identify is to question, to turn all witnesses into subjects of a particular question: 'since this is nature, what should be done to preserve it?' In sum, we must recognize the pressure of the environmental horizon towards the introduction and maintenance of a paradigm that seeks to put nature above – rather than within – society in order to generate conscience, options and decisions where there were previously none.

The consequences of restructuring the relationship with nature throughout all identifiable fields of action and conscience should not therefore be dismissed as a mere osmosis between the natural and the social. On the contrary, we must try to understand how the new perceptions compete with older ones and look at the dynamics of that antagonism. At this point an intriguing general picture of the alternative model for power that is inherent in the environmental claim may be drawn: firstly, because power is an all-embracing category, cutting through the social, yet identifiable on its own account; secondly, because we cannot assert that the environmental horizon aims at preserving nature, since there can be no consensus on what constitutes nature or the preservation of nature. With the same perfect plausibility, however, one may argue that the environmental horizon aims to modify the equilibrium of social forces towards a new model of setting and administering priorities in human societies. I shall briefly outline below some of the main orientations of this emerging proposal of power management and distribution.

Totality Replaces Individuality

This is a direct consequence of the fact that environmental perceptions and practices do not make use of any metaphysical or symbolic mediation in order to represent nature in society but rather, directly integrate society into nature. Obviously then, human society itself has to carry out all necessary action in order to conserve a specific natural condition. After all, there is no other regulator to guarantee or justify the continuation of this socially integrated natural world. Thus, the socio-political replaced the metaphysical, and responsibility lies now on the shoulders of human societies and their voluntary priorities. A constant reference to nature cannot be insured without involving a conscious priority based on the very fact that nature is plain existence. It has itself no point of view but corresponds to human points of view and, most importantly, is crucially subjected to the priorities of human societies.

As a result, a coherent and collective scheme must necessarily preserve and cultivate the kind of social action that submits itself to the overarching, allegedly exogenous, priority of the natural environment. Social antagonisms are then to be seen not only as an issue of social justice legitimated by conflict of interests within society, but also as a space of conscience beyond the social universe as such. It is not enough for the individual or the institution to pretend that their relations to other agents legitimate their action. What is also required is that they align their interaction to the external priority given to other species, and that they plan, control and realize their action in society not only with regard to their competitive sociocentric interests but also with regard to the environment. Otherwise put, it is not enough to be right in terms of social conflict in order to demarcate a particular course of events and try to bring it into existence; it is also indispensable that such a course of events reduce (or at least, does not increase) its impact on 'nature'. If this cannot be achieved, it is preferable, following the Socratic reasoning, to suffer

Energy and Culture

injustice rather than to cause it, particularly against some entity as defenceless as nature, which cannot even be symbolically represented. It is difficult to see how nature could be symbolically represented in a unified way, particularly since its representations reflect social positions and range from eternal stability to urgent fragility; all work conducted under the Grid/Group methodology is highly enlightening here (e.g. Schwarz and Thompson, 1990; Thompson et al., 1990; Ellis and Thompson, 1997; Douglas, 1993). In general, it is true that both the influence of environmental campaigns and the progressive understanding of the earth as a 'small planet' have in the last thirty years increasingly shifted the social conscience of nature towards fragility.

In that context, social interaction is becoming a matter of mediation and negotiation towards the end of supervising a harmony between all participants of the natural world. By the same token, individual or partial interests are to be assessed in the light of this harmonious relationship (see Twine, 1994) and should give way to the modalities of that relationship, even if that entails suffering unjust loss when one thinks exclusively in terms of a (human) social context. This is even more the case if we deal with actions that are already a product of domination. For it is obviously far less tolerated to produce negative consequences in regard to the environment when one simply seeks to increase one's power within society than it is when one struggles for survival. Since the latter is already under scrutiny in an environmental context of judgement and cannot occupy any other position than that of the victim, the former becomes by contrast an actor driven by utmost cruelty and indifference both towards the social and the natural environment.

Accordingly, a general framework emerges, which is meant to determine the priorities and limits of individual action. Power is no longer to be produced and distributed in terms of the capacity to realize a project of action through social relations; under the environmental horizon, it can and should be produced, managed and increasingly distributed within a total world view that negates individual motivation and develops as an *interface* between human societies and their environment.

Equilibrium Replaces Order

The second major transition that arises due to 'the environmental horizon' has to do with the prescriptive dimension of power. In fact, this dimension appears so self-evident as to have been taken for granted, even by social scientists. Not only do we not need a reason to justify the existence of power as such, but we are also ready to accept that power's inherent end is to structure the world in the particular fashion that increases the benefits of those who exert it. The environmental horizon, however, seriously challenges the self-evidence of the relationship between power and its ends. This is due, on the one hand, to the very fact that human domination over other species and nature in general is widely accepted as a given truth. Collective human power is not put in doubt *as* a reality but *because* it constitutes such an inevitable reality and can therefore shape the world according to desires and priorities peculiar to human societies. Today, this means shaping the world according to the desires and priorities of those groups that are dominant in late industrial societies. *Problematory pursuit of power*

On the other hand, reference to spheres which are exterior to human societies, such as other species, 'the environment', 'nature', etc., begs directly the question of the meaning of power as such. For if it is legitimate to think that in human interaction seeking power is simply natural, even if this is only because almost everyone else does so, it is impossible reliably to answer the question of whether it is natural, legitimate or otherwise justified to seek to dominate and transform the world of other beings and nature in general. That can only be a matter of judgement; and in being so it casts a serious doubt over the use of power. Thus, the environmental horizon pushes power out of a prescriptive register into a functional one. This is why political opinions are excellent predictors of public attitudes on the environment (Papadakis, 2000). Submitting the use of power to functional criteria and concrete outcomes, such as easily quantifiable pollutant rates, is in itself a critical understanding of market-led growth, particularly when it is proposed in the interest of other beings rather than humans themselves. Two major consequences result from this transition. The first one is that power is no longer a natural condition; it has specific functions and a utility which its exertion must practically prove. In other words, there is a scale for assessing power with 'prescriptive'/'autistic' at the bottom and 'creative'/'sharing' at the top. The second consequence is that power cannot be discussed without considering the extent to which it serves the relationship between human societies and their environment. In that sense, this relationship becomes a permeating, omnipresent control factor for the exercise of power in the environmentally aware society.

These two perspectives can be combined into a unified view based on the distinction between order and equilibrium. The environmental horizon suggests that the natural condition is that of seeking harmony and balance rather than deciding on the grounds of convenience, and in doing so it introduces a new role for the socio-political register. This role has more to do with achieving a state of permanent (i.e. 'sustainable') homeostasis, both in society and in nature, rather than with the notion of producing a comfortable order of things by using all possible means. *— Susty as homeostasis*

Doubt Replaces Belief

Seen from an environmental point of view, 'nature' is neither utilitarian nor neutral; and in not being so it should not be approached through aesthetics. There are undeniably several ways of observing this rule; and they correspond to different degrees of resistance to a perception of the natural environment as 'heritage' or as source of human well-being. 'Deep' environmentalism is different from an 'ethical' New Age approach, which again has little to do with the scientific-technocratic discourse on internalizing environmental cost, or with the 'apocalyptic' environmental perspective (for a comparative theoretical discussion of different perspectives on environmental reform see Spaargaren and Mol, 1993). Such great disparities are socially and politically meaningful and should not be glossed over. On the contrary, by taking into account the diversity of these

deep green vs technocratic internalizing / CBA.

basic principles of environmental horizon.

attitudes, the significance of the priorities that they have in common can be revealed fully. Although there is more than one way in which consensus can materialize, one can plausibly argue that some basic principles underlie the whole spectrum of the environmental horizon. First and foremost, it is not possible to develop any interest – and consequently any discourse – on environmental protection without *worrying* about the environment. An environmental perspective presupposes that there are problems to be researched and resolved. In other words, a condition of fragility and threat is projected on the natural world and reflected back to social representations. It is possible to associate the necessity of that process with a series of interdependent conditions: *fear* in terms of an individual link to experience; *threat* with regard to the content of the surrounding world which is no longer stable or reassuring; *uncertainty* in terms of constituting a sustainable hypothesis concerning the continuity of the environment; *prevention* with regard to the operational and decisional aspects of acting on an individual, collective and institutional level. *fear + threat as basis for action*

The prevention model turns fear, threat and uncertainty into a basis for action rather than inertia (Lascoumes, 1996). It also reverses one of modernity's best-established premises of action, i.e. that it is legitimate to pursue a course of action unless it is certain that such a course will produce negative consequences for human societies (or, more accurately, for the powerful parts of some human societies). Precautionary management establishes a model of decision-making, which integrates uncertainty as a sufficient cause for defensive action (or abstention from action) when there is a suspicion that a threat would intensify or materialize. Doubt is, therefore, increasingly seen as a normal decisional guide while certainty is being undermined as a basis for proceeding with well-informed options, and appears increasingly as *naiveté*. This is why today one can increasingly find supporters of the idea that in the post-industrial societies those who are still happy are only those who have not yet heard the bad news or who do not want to listen to it. In this context, understanding the world becomes an issue of competence in developing hypotheses about it, rather than a matter of adhering to fixed schemes of knowledge and action in order to approach it. *competing with belief*

From this point of view the environmental horizon competes with belief *per se* as a valid means both of producing meaning and of managing the environment, and in doing so it challenges the legitimacy of claims to cognitive stability. Again, this is not to say that expression of clear political conflict is not possible within the environmental horizon. For example, it can be plausibly argued that the distinction between Right and Left persists (Scott, 1990). There are also attempts to generate radical environmentalism, for example in the form of eco-socialism (Bond et al., 2000). In particular, integrating doubt on an operational level practically excludes all recourse to prescriptive systems of thought and action and sets the base for a cultural edifice which tolerates only one principle, i.e. that only commitment to a cause is useful, while all kinds of limits that may stem from prearranged belief systems (individual or otherwise) are essentially inefficient as tools of understanding and managing the world. In fact, within the environmental cultural horizon, the world around us is precisely what holds the keys to steering the appropriate course of action; and monitoring our relationship with that world is the

Precautionary Principle

undermining certainty

doubt as normal challenges stability

eco-socialism

excluding prescriptivism?

only source of reliable knowledge and decision-making (Berque in Bourg, 1993). In this context, *doubt becomes for all practical purposes a commitment*, both individual and collective, *to diligence and openness* in our interaction with the external world.

It is not simple to decipher what these developments mean with regard to power. However, there is some safe ground on which to build, as far as institutions are concerned. Managing a context of action – that is, carrying out a project – has historically been an issue of conceiving, programming and applying specific initiatives; according to the power structures that are still largely operating in post-industrial societies, firstly the natural and then the social world were there to guide the humans towards an anthropocentric, utilitarian and, needless to say, class-defined perspective on the future. Introducing doubt radically changes that configuration, since the suggestion is now that the truth on the preferable course of action is held not by the decision-making system but by the environment to which decisions are to be applied. At the same there is widespread acceptance that this environment is a partially unknown factor on which one hypothesis can often be as valid as another. Power is being therefore mediated by its own object and subjected to a wide conscience legitimizing access to decision-making for all social participants. What is at stake here is authority as such, either based on specialized knowledge (e.g. scientific expertise) or on institutional hierarchical position (see van Koppen, 2002). Those who 'know better' or 'are in charge' are thus exposed to scrutiny based on criteria that are beyond their control and reflect the wishes, rights and capacities of lay people. There is serious thinking and rigorous research on integrating public participation down to the level of choices in scientific methodology, and on the consequent acceptability of research (e.g. Ravetz and Funtowicz, 1999; Marris, 1999; Marris and Joly, 1999). Moreover, the breadth of issues for which there is legitimacy of lay opinions extends to almost every issue, since there is by definition no action without environmental impact. The environmental horizon thus suggests that some of us may know more about the problem and a possible solution, but at the same time it denies any priority to a particular group or class with regard to defining which problem is more important and which solution should be followed. Based on doubt, it introduces a *de facto* participative model of distributing power and managing both nature and society. This is also why environmentalism is being used as a vehicle for promoting democratic change within societies that have weak structures of political participation (Kim, 1999) and, increasingly, as a symbolic axis for ethnic, social or political conflict (Timura, 2001).

Dangerisation and Fluidity

If one is prepared to accept that the introduction of totality, equilibrium and doubt through the environmental horizon restructures both the possession of power and its object, one is bound to ask what makes this transformation coherent in cultural terms. We witness a reversal of values, means and methods for approaching the relationship between understanding and managing our environment or, put in temporal terms, between the present and the future. In fact all projects of action in

[handwritten top margin: how about top-down means to overcome dangers – we've many beyond the debating phase!]

the post-industrial society converge towards the detection of possible dangers both as an instrument of knowledge and as a basis for judgement. The environmental horizon builds an image of the natural and the social world by looking at them as sources and objects of danger at the same time. That process of dangerisation is hardly neutral and entails a conflict with the technocratic model of administering resources. The concept of dangerisation addresses critically the post-industrial obsession with risk and looks at the 'risk society' as an epiphenomenon of deeper socioeconomic change. It focuses on the decline of sociality and politicization as the causes that make risk a convenient fulcrum for late modern power structures (Lianos, 2001; Lianos and Douglas, 2000; Lianos, 1999). Beyond risk management – that is now an integrated part of contemporary decision-making – dangerisation is the cultural process of extending the conscience of threats onto the life-world and of imposing on the socialised individual, both as a right and an obligation, the task of searching for possible threats and developing efficient strategies in response. This trend in conscience and experience is highly disruptive for the technocratic model, which concentrates on uniform, uninterrupted processes in order to secure continuous flows between the interacting components of the social and the natural universe. While technocratic management only allows for questions that are subsequent to the objective of organisational fluidity, the environmental horizon introduces issues of principle as a legitimate reaction to a dangerised world. The administrative question 'how can we diminish the impact of factory X on the environment?' is now in competition with its environmental counterpart: 'should factory X exist in the first place?'.

[handwritten left margin: I'm not so sure about this! risk as convenient term? can also extend behaviour]

[handwritten below paragraph: matter for policy – case by case; poor grasp of behaviour, I'd say]

Environmentalism, Power, Uncertainty: A Concluding Question

The environmental horizon is deeply challenging both for the perception and the utility of power in human societies. There is surely a long-term effect in that process which cannot be accurately foreseen but seems nonetheless to be large-scale. This effect involves the 'taming' of power as such in the context of human societies, and its transformation from a tool and a consequence of domination into a conscious organizational function. One can hardly imagine a more ambitious plan for human societies. However, in the short and medium term it is, I think, quite certain that the social development of the culture involved in this process will lead to an increased loss of points of reference: firstly, with regard to conflicts on defining the priorities of the transition described here; secondly, with regard to the fragility of the social and the natural world. *[handwritten: disempowering effect]*

[handwritten left margin: Disruptive 'taming' of power]

There are many reasons to believe that a spreading conscience of risk and insecurity deprives post-industrial societies of a great part of their potential for change and leads to a neutral conservatism, more appropriate for fear and defensiveness than for the emergence of challenging social and political attitudes. In that sense, adding the element of doubt concerning both the context and the meaning of power endangers the stability of those cultural formations that historically have founded social organization, at least from modernity onwards. This mutation is probably useful and in fact welcome, but it also underlines the fact that restructuring the social and the political can only take place at a serious cost.

[handwritten bottom: costs of env't perspective]

CONCLN

In this case, *the environmental horizon attempts to produce change in the power structure of post-industrial societies at the price of increasing insecurity.* The price for eliminating environmental uncertainty is the surge in danger awareness both at an individual and a collective level.

This process is neither conscious nor intended. Like all major social transitions, it stems from the limitations of a world whose sociopolitical organization cannot catch up with its level of conscience. Like all major social transitions, it represents a solution to a series of problems that seemed unrelated until the change is almost complete. Like all major social transitions, it builds on the formation of a radically new conscience: we are to see ourselves as the victims of our own actions, the repenting, tragic subjects of our hubris whose catharsis we should ourselves instigate. There lies also the uniqueness of the environmental claim: the subjects of change are not the innocent, the poor, the weak, but the wealthy and the powerful. This is also true in developing countries, where the calls for environmental preservation come from the educated middle classes (e.g. Munshi, 2000). We are not asked to think of ourselves as the exploited but as the exploiters. We are expected to act out of guilt, not out of hope. What Christianity did for other human beings, the environmental perspective does now for other animals and plants. Human self-discipline is at the centre of the project. Fear of God is now replaced by fear of a major accident, a rapid natural change that will eliminate, decimate, or mutate our species. At least, one could say, we have progressed: these fears correspond to something observable. But, on the other hand, what is always there is the conscience of our primordial importance as a species whose domination of the planet should know no end, which is what makes power so central to our culture. Perhaps the next narrative of social transformation will appeal to another level of consciousness, where the power structure will have to accommodate an unleashed massive hedonism, suggesting that the history of our species should end at some point. But for the time being, we will have to content ourselves with exploring the social limits of self-preservation.

speculates on end of our species – prefer hedonism rather than eventual susty giving up on susty

References

Allan-Michaud, D. (1989), *L'avenir de la société alternative: les idées 1968-1990 ...*, L'Harmattan, Paris.

Beck, U., Giddens, A. and Lash, S. (1994), *Reflexive Modernization: Politics, Tradition and Aesthetics in the Modern Social Order*, Polity Press, Cambridge.

Bond, P., Miller, D. and Ruiters, G. (2000), 'Regionalism, Environment and Southern African Class Struggles', *Capitalism, Nature, Socialism*, vol. 11, no 3.

Bourg, D. (dir.) (1993), *La nature en politique – ou l'enjeu philosophique de l'écologie*, L'Harmattan, Paris.

Castree, N. (2002), 'Environmental Issues: From Policy to Political Economy', *Progress in Human Geography*, vol. 26, no 3.

Caillé, A. (2001), 'Une politique de la nature sans politique: à propos des politiques de la nature de Bruno Latour', *la revue du MAUSS*, vol. 17.

Douglas, M. (1993), 'Governability: A Question Of Culture', *Millennium*, vol. 22, no 3.

Ellis, R.J. and Thompson, F. (1997), 'Culture and the Environment in the Pacific Northwest', *American Political Science Review*, vol. 91, no 4.

Elliott, E., Regens, J.L. and Seldon, B.J. (1995), 'Exploring Variation in Public Support for Environmental Protection', *Social Science Quarterly*, vol. 76, no 1.

Jonas, H. (1984), *The Imperative of Responsibility: in Search of an Ethics for a Technological Age*, University of Chicago Press, Chicago.

Kemmelmeier, M., Krol, G. and Kim, Y.H. (2002), 'Values, Economics, and Proenvironmental Attitudes in 22 Societies', *Cross-Cultural Research*, vol. 36, no 3.

Kim, D.S. (1999), 'Environmentalism in Developing Countries and the Case of a Large Korean City', *Social Science Quarterly*, vol. 80, no 4.

van Koppen, C.S.A. (2002), 'Environmental Discourse and the State: A Social Analysis of Debates on Transport and Environment in Portugal and the Netherlands', *Research in Social Problems and Public Policy*, no 10.

Lascoumes, P. (1994), *L'éco-pouvoir: environnements et politique*, Paris: La Découverte.

Lascoumes, P. (1996), 'La précaution comme anticipation des risques résiduels et hybridation de la resposabilité', *Année Sociologique*, no 2, vol. 46.

Lianos, M. (2001), *Le nouveau contrôle social: toile institutionnelle, normativité et lien social*. L'Harmattan-Logiques Sociales, Paris.

Lianos, M. and Douglas, M. (1999), 'Point de vue sur l'acceptabilité sociale du discours du risque', *Les Cahiers de la Sécurité Intérieure*, 'Risque et démocratie', no 38.

Lianos, M. and Douglas, M. (2000), 'Dangerization and the End of Deviance: The Institutional Environment', special edition of the *British Journal of Criminology*, vol. 40, April, 2000; also in Garland, D. and Sparks, R. (eds), *Criminology and Social Theory*, Oxford University Press, Oxford.

Marris, C. (1999), 'GMOs: Analysing the Risks', *Biofutur*, no 195.

Marris, C. and Joly, P.B. (1999), 'Between Consensus and Citizens: Public Participation in Technology Assessment in France', *Science Studies*, vol. 12, no 2.

Martell, L. (1994), *Ecology and Society: An Introduction*. Polity Press, Cambridge.

Milbrath, L., Downes, Y. and Miller, K. (1994), 'Sustainable Living: Framework of an Ecosystemically Grounded Political Theory', *Environmental Politics*, vol. 3, no 3.

Munshi, I. (2000), '"Environment" in Sociological Theory', *Sociological Bulletin*, vol. 49, no 2.

Papadakis, E. (2000), 'Environmental Values and Political Action', *Journal of Sociology*, vol. 36, no 1.

Ravetz, J.R. and Funtowicz, S. (1999), 'Post-Normal Science: An Insight now Maturing', *Futures*, vol. 31, no 7.

Saurin, J. (2001), 'Global Environmental Crisis as the "Disaster Triumphant": The Private Capture of Public Goods', *Environmental Politics*, vol. 10, no 4.

Schwarz, M. and Thompson, M. (1990), *Divided We Stand: Redefining politics, Technology and Social Choice*. Harvester Wheatsheaf, Hemel Hempstead.

Scott, A. (1990), *Ideology and the New Social Movements*. Unwin Hyman, London.

Smith, M.J. (ed.) (1999), *Thinking through the Environment: A Reader*. Routledge, London.

Spaargaren, G. and Mol, A.P.J. (1993), 'Environment, Modernity and the Risk Society: the Apocalyptic Horizon of Environmental Reform', *International Sociology*, no 4.

Sperber, D. (1996), *La contagion des idées: théorie naturaliste de la culture*. Jacob, Odile Paris.

Thompson, M., Ellis, R. and Wildavsky, A. (1990), *Cultural Theory*, Boulder. Westview Press, CO.

Timura, C.T. (2001), 'Environmental Conflict and the Social Life of Environmental Security Discourse', *Anthropological Quarterly*, vol. 74, no 3.

Twine, F. (1994), *Citizenship and Social Rights: The Interdependence of Self and Society*, Sage, London.

Weaver, A.A. (2002), 'Determinants of Environmental Attitudes', *International Journal of Sociology*, vol. 32, no 1.

Wynne, B. (1987), *Risk Management and Hazardous Waste: Implementation and the Dialectics of Credibility*. Springer, Berlin.

Chapter 13

Energy Technologies and Integrated Risks

Natasa Markovska, Nada Pop-Jordanova and Jordan Pop-Jordanov

[handwritten: Energy Studies + Healthcare Macedonia]

Energy is an essential input for social development and economic growth. It provides for basic needs such as heating, cooling, lighting and transport and it is also a critical production factor in practically all sectors of industry. However energy generation and energy use induce a variety of environmental and health risks. Depletion of natural resources, accumulation of wastes, deforestation, and territorial devastation are all energy-related threats to future generations, jeopardizing sustainable development.

Risk and Sustainability: Basic Models

[handwritten: Correlating Risk + Sustainability]

One of the most important problems we face is how to supply enough energy in a sustainable way, i.e. without unacceptable damage to health and the environment and without compromising the ability of future generations to meet their own needs. In this respect, an approach correlating risk science and sustainability science can be used to improve the safety of energy generation and use, as well as energy policies and decision-making. The following sections will attempt to apply this correlative approach to energy technologies by presenting the basic models of both risk science and sustainability science. In risk science, the basic model is the so-called NAS Risk Paradigm, released by the US National Academy of Sciences (National Research Council, 1983). In sustainability science, the basic model is the Pressure-State-Response (PSR) model (National Research Council, 1999), applied for compilation of numerous sets of Indicators for Sustainable Development (ISD).

NAS Risk Paradigm

[handwritten: Using models developed in US]

The paradigm proposed by the US NAS provides a common framework for addressing many kinds of risks, including those related to energy technologies. As presented in Figure 13.1, this scheme defines Risk Assessment as a four-step process, operationally independent from Risk Management, including hazard identification (1), dose-response (2) and exposure (3) assessment, and risk characterization (4).

[handwritten: A step model, rather functional]

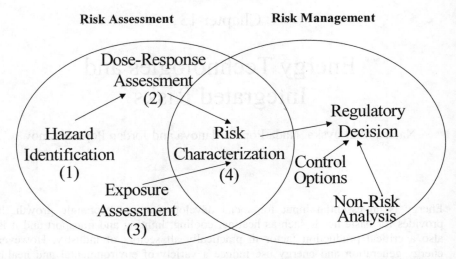

Figure 13.1   The NAS risk paradigm (National Research Council, 1983)

The questions addressed at each step of the Risk Assessment process are as follows: (1) Is this substance toxic? (2) How toxic is it? (3) Who is exposed to this substance, how long, how often? (?) (4) So what? The answers to these questions represent risk information as one of several factors that the decision-maker must consider when defining Risk Management activities.

Based on the NAS risk paradigm, typical results from comparative health risk assessments of different energy technologies (performed by several authors) are presented in Figure 13.2. The health risk is expressed in man-days lost per megawatt-year or in fatalities, assuming that one fatality is equal to 6000 man-days lost due to injury or disease. When determining the health risk values, all stages of plant life and fuel cycle are included, from plant construction to dismantling and from mining to waste disposal. The highest health risks are associated with coal plants, predominantly taking the form of occupational riskin the case of coal mining and of public risk connected with atmospheric pollution in the case of energy generation. The same holds true for oil plants, with more emphasis on public risk. The lowest level of health risk is identified for gas and nuclear plants, owing to the low emission of pollutants in the case of gas and to the high energy density of fuel in the case of nuclear power. A relatively high level of health risk is associated with hydropower due to frequent accidents with dams, and also with the solar technologies due to the large amount of material required in the pre-operational stage and the need for an intermittent backup energy system.

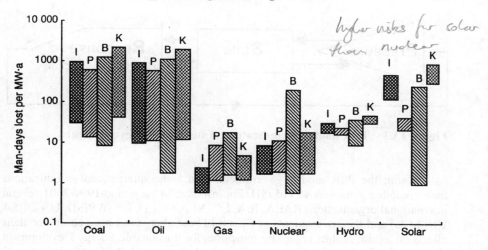

[handwritten annotation: higher risks for solar than nuclear]

Figure 13.2 Standard health risk values for different energy technologies

Source: Pop-Jordanova and Pop-Jordanov, 1996

Note: I-Inhaber, P-Pop-Jordanov, B-Brikhofer, K-Kollas and Papazoglou *[handwritten: 4 different studies]*

However, the assessment of the health risks involved in energy technologies involves a rather high degree of uncertainty. As can be seen from Figure 13.2, the calculated health risk values have relatively large bandwidths, suggesting the range of uncertainties and variations among different data sets. *[handwritten: – and big difference in uncertainty for same technology across diff studies]*

The PSR Model for ISD

Indicators of Sustainable Development (ISD) are defined as quantifiable parameters for measuring and monitoring changes (i.e., progress or degradation) with respect to sustainable development, thus signalling challenges or alarms. Generally, they provide a quantitative index covering economy, social well-being, and impact of human activities on the natural world.

The current sets of ISD are built upon the Pressure-State-Response (PSR) model, which, as shown in Figure 13.3, incorporates three types of interacting variables. A human activity exerts pressure that changes the state variable, which in turn should evoke response. The last can act in two ways: in some cases it acts on the state variable, which can further affect the pressure; in other cases it can directly control or correct the pressure.

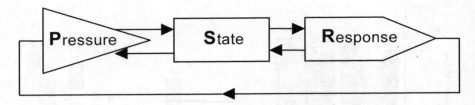

Figure 13.3   PSR model for indicators of sustainable development

Using the PSR model, governments and some international organizations have compiled numerous sets of ISD. For instance, in the period 1999-2001 several international organizations (IAEA, IEA, EC, NEA/OECD, UN-WPISD, UN-DESA, UNESCO) and the governments of fifteen EU member states have combined their efforts on a joint project regarding Indicators for Sustainable Energy Development (ISED). The final result of this project is the full set of 41 ISED, of which the 23 listed below are identified as the core set of ISED (IAEA-IEA, 2001):

1. End-use energy prices with and without tax/subsidy
2. Energy intensity: manufacturing, transportation, agriculture, commercial and public services, residential sector
3. Energy mix: final energy, electricity generation, primary energy supply
4. Energy supply efficiency: fossil fuel efficiency for electricity generation
5. Energy use per unit of GDP
6. Expenditure on energy sector: total investments, environmental control, hydrocarbon exploration and development, RD&D, net energy import expenses
7. Energy consumption per capita
8. Indigenous energy production
9. Net energy import dependence
10. Fraction of disposable income/private consumption spent on fuel and electricity by: a) average population; b) group of 20 percent poorest population
11. Fraction of households: heavily dependent on non-commercial energy; without electricity
12. Quantities of air pollutant emissions (SO_2; NO_x particulates, CO_2, VOC)
13. Ambient concentration of pollutants in urban areas: SO_2, NO_x, suspended particulates, ozone
14. Quantities of greenhouse gas emissions
15. Generation of solid waste
16. Accumulated quantity of solid wastes to be managed
17. Generation of radioactive waste
18. Accumulated quantity of radioactive wastes awaiting disposal
19. Land area taken up by energy facilities and infrastructure
20. Fatalities due to accidents, with breakdown by fuel chains

21. Fraction of technically exploitable capability of hydropower currently in use
22. Proven recoverable fossil fuel reserves
23. Intensity of use of forest resources as fuel wood

Analyzing the list, one may easily conclude that the current ISED take account of material and energy resources and their impacts, with no concern for mental resources and capacities employed in energy production and use.

[handwritten: Curious comment on 'mental resources']

Risk Related ISED

In order to apply a correlative approach to energy technologies, incorporating risk and sustainability, we introduce the Risk Related ISED (RR ISED), including indicators from the core set regarding three different types of risks related to air pollutants, wastes, and fatalities. The following is the corresponding classification:

Air pollutants:
12. Quantities of air pollutant emissions (SO_2, NO_x, particulates, CO_2, VOC)
13. Ambient concentration of pollutants in urban areas: SO_2, NO_x, suspended particulates, ozone
14. Quantities of greenhouse gas emissions

Wastes:
15. Generation of solid waste
16. Accumulated quantity of solid wastes to be managed
17. Generation of radioactive waste
18. Accumulated quantity of radioactive wastes awaiting disposal

Fatalities:
20. Fatalities due to accidents, with breakdown by fuel chains

According to this analysis, eight out of twenty-three ISED from the core set are clearly risk-related. Further analyses of the mentioned IAEA/IEA core set (and also the full set) have shown two weak points. The first weak point concerns the third group of RR ISED, wherein only risk induced by accidents is taken into consideration, neglecting the risk due to the normal operation of the plant. This specific gap could be easily filled, as is done in Figure 13.1. However, the other weak point is of more general and conceptual nature. As we shall show in the next section, it consists in overlooking occupational entropy (Pop-Jordanov, Markovska et al., 2004) and the corresponding indicators that should account for the threats to mental capacities and resources. The additional group of RR ISED, which we call Mental Indicators (Pop-Jordanov, 2003) could provide for a proper treatment of this kind of 'mental risk' associated with energy generation and use.

[handwritten: "occupational entropy"? (mental risk) ??]

Occupational Entropy

The world of today is characterized by a progressive shift from energy and material resources to mental ones. Indeed the role of mental labour has become such that we may now speak of a knowledge-based economy. However, if we consider all aspects of mental labour, our attention should be directed particularly to a 'wisdom'- or even 'mind'-based economy (Pop-Jordanov, 2003). According to the UNESCO definition, a wisdom-based economy would signify an economy built upon knowledge and morality. Furthermore, all mental dimensions are taken into consideration when one talks about a mind-based economy, conceiving mind as a totality of cognition, emotion and morality.

Under these circumstances, the increasing entropy within the world economic system ceases to be attributable only to physical variables, but more and more includes threats to mental resources and capacities. The new concept of occupational or mental entropy, therefore, addresses disorders and degradation caused by cognitive, emotional and moral agents (Pop-Jordanov, Markovska et al., 2004).

Information Overflow and Organizational Attention Deficit

Due mainly to the surplus of daily messages and data, the modern information age, besides offering advantages, brings about a variety of unfavourable situations, such as the following:

- An increased likelihood of missing key information when making decisions;
- Diminished time for reflection on anything but simple information transactions such as e-mail and voice mail;
- Difficulty in holding others' attention;
- Decreased ability to focus when necessary.

All these occurrences are symptoms of the so-called organizational attention deficit disorder (Davenport and Beck, 2001). The adverse effect of information overflow is exacerbated by other factors, such as psychological conditions related to the working environment and the kind of job, which also can reduce occupational attention and concentration.

In the energy sector, induced attention deficit may have a number of detrimental consequences. An extreme case is the Chernobyl accident, in which the human factor proved to be a crucial one. Seen from a more general perspective, attention has a progressively enhanced role in modern economies. In some contemporary economic theories, attention is indicated as a new currency of business (Davenport and Beck, 2001).

Considering the threat to cognitive capacities, including attention, posed by information overflow, this clearly deserves to be taken seriously in order to mitigate its harmful influences.

Occupational Stress and Psychosomatic Diseases

In parallel with physical/chemical pollution, energy technologies may be sources of 'psychological pollution' with analogous consequences for human health and welfare. The actual or potential technological hazards associated with energy production may provoke considerable emotional disquiet expressed in different forms of stress.

In particular, the presence of psychological pollution may be confirmed if we compare opinions and facts about the health impact of some energy technology. To this end, the medical staff for health protection near the Bitola coal-fuelled power plant was surveyed on their opinions concerning a possible increase in the number of respiratory diseases caused by plant operation (Figure 13.4).

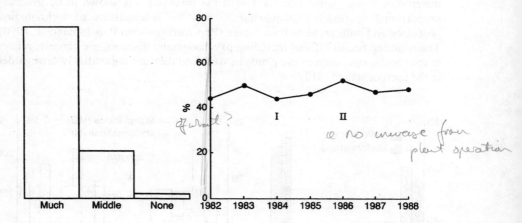

of what?

& no increase from plant operation

Opinions of the medical staff about the influence of the plant on respiratory diseases

Factual incidence of respiratory diseases before and after the introduction of plant units I and II

Figure 13.4   Chronic emotional stress induced by the coal-fuelled power plant Bitola

Source: Pop-Jordanova and Pop-Jordanov, 1996

As can be seen, almost 80 percent answered that the number of diseases had greatly increased, compared with less then 2 percent of the surveyed staff who thought that there was no relation between the number of respiratory disorders and plant operation. Contrary to the prevailing opinion of the surveyed medical staff, the factual incidence of respiratory diseases, evidenced before and after installation of plant units I and II, did not change (Figure 13.4). Evidently, this large discrepancy between opinions and facts derives from the presence of psychological pollution associated with plant operation.

not substantiated by this shaky evidence

"psychological poll'n" bizarre.

Similar conclusions can be drawn from a study comparing actual malignancies and malformations among children near Skopje after the Chernobyl accident to opinions concerning these effects (Figure 13.5). The medical staff of the Paediatric Clinic (89 medical doctors and 50 nurses) was surveyed and the results of the survey were compared to the factual incidence. Again an exaggerated discrepancy between opinions and facts was shown, illustrating the presence of psychological pollution among many people, including professionals.

Mostly, induced stress represents a chronic emotional disturbance. Refracted through the personality prism, which comprises heredity, early child experiences and actual conflict, it may lead to a spectrum of psycho-physiological disturbances and psychosomatic diseases. The last are induced following the inverse causality model in risk assessment: internal (psychological) agent → functional disorder → material or tissue harm. The level of stress involved was shown to be inversely proportional to people's acquaintance with the mechanisms of technological processes and with protection measures (Pop-Jordanov and Pop-Jordanova, 1990). The resulting health effects, including psychosomatic diseases, are chronic, related to the normal operation of the plants as well, and they are unjustifiably disregarded in the present sets of ISED. ← *make the case for more explicit recognition of psychological harm of nuclear power*

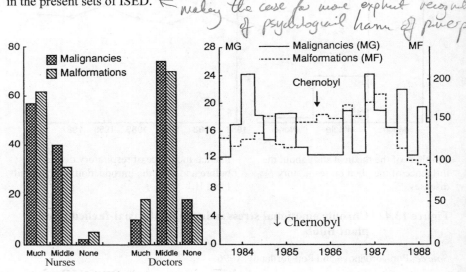

Opinions of paediatric staff in Skopje concerning the increase of malignancies/malformations of children in the region after the Chernobyl accident

Factual incidence of malignancies and malformations of children in the Skopje region before and after the Chernobyl accident

Figure 13.5   Chronic emotional stress induced by low doses of nuclear radiation

Source: Pop-Jordanova and Pop-Jordanov, 1996

Corruption Pressure and Reduced Returns

Additional harmful pressure on mental resources comes from the rising incidence of corruption and other deviations characteristic of economies in transition. Acting destructively on moral capacities, as well as on the whole of economic and social life, these phenomena are reflected in various forms of reduced returns, such as detrimental contracts, improper employment or loss of human capital.

This kind of pressure is of particular importance in the energy sector, with ethics and transparency being preconditions for conducting an efficient and timely restructuring and privatization.

[handwritten: mental stress of corruption in transition economies]

Risk Related Mental ISED

[handwritten: Indicators of sust'ble energy development are unable to address mental risks]

The current RR ISED quantify issues related only to material and energy resources and their impacts, omitting issues regarding mental resources and capacities involved in the processes of energy generation and use. In other words, occupational entropy is disregarded, making the current Indicators for Sustainable Energy Development unable to address the mental risks associated with energy technologies. This defect can be corrected by proper extension of the current sets of indicators by introducing a new group of RR ISED, namely Mental Indicators, which correspond to types of mental resources and capacities, and include cognitive, emotional and moral indicators. In this way, one may talk about integrated risks from energy technologies, which besides the standard ones include also mental risks.

For the sake of consistency, the newly introduced RR Mental Indicators are represented in the standard PSR framework.

Cognitive Indicators

PSR representation of the cognitive indicators is illustrated in Figure 13.6. The pressure variable of the cognitive indicators is information overflow. As described in the previous section, this provokes attention deficit, reflected in the incidence of organizational attention deficit disorder. A possible response is to introduce attention strengthening techniques, including biofeedback training.

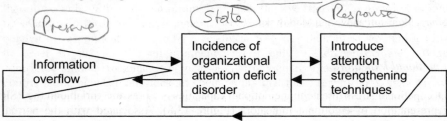

Figure 13.6   PSR model for RR cognitive ISED

Biofeedback

Biofeedback could be succinctly defined as electronic measuring, processing and feeding back of information about one's own inner activity, to help one achieve voluntary control of symptoms (e.g. attention deficit or stress). Depending on the parameter targeted by biofeedback, this technique has different modalities. Attention improvement in the form of cognitive capacity is effected by adjustment of brain wave frequencies using the electroencephalographic modality, i.e. EEG biofeedback training.

In terms of brain wave frequencies, attention deficit disorder is manifested in the increased theta/beta ratio of brain waves. Therefore, the final goal of EEG biofeedback treatment is to achieve reduction of this parameter. As an example, in Figure 13.7 the results of EEG biofeedback treatment, including 40 sessions of 12 patients, are presented, giving the values of the mean theta/beta ratio for each *brain wave* patient, before and after treatment. As can be seen, considerable decrease of this *scans* parameter is achieved, which, along with improvements in clinical symptoms, proves the efficiency of EEG biofeedback as an aid in strengthening attention.

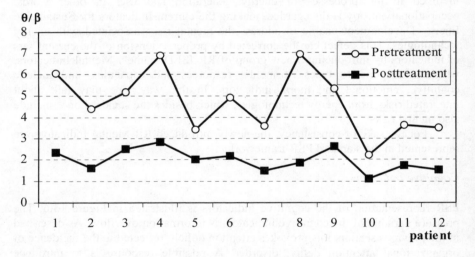

Figure 13.7   Mean theta/beta ratio as attention deficit indicator

Source: Pop-Jordanova and Markovska, 2001

Emotional Indicators

Occupational stress threatening emotional capacities is a pressure variable in the PSR representation of emotional indicators (Figure 13.8). Associated with the normal operation of power plants, its permanent presence may cause psychosomatic diseases, such as peptic ulcer, arterial hypertension, colitis, neurodermatitis and asthma. In this sense, the incidence of psychosomatic diseases can be considered as a state variable.

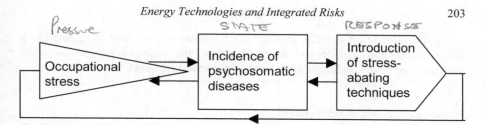

Pressure   STATE   RESPONSE

Figure 13.8   PSR model for RR emotional ISED

Some relevant studies conducted among the staff of the Macedonian coal power plant at Bitola, and the Slovenian nuclear power plant at Krsko confirmed the existence of chronic stress due to occupational hazards (Figure 13.9). Namely, for both plants there is a considerable difference in the incidence of various psychosomatic diseases (expressed as a percentage of total number of patients in the group) between the plant staff and the control group.

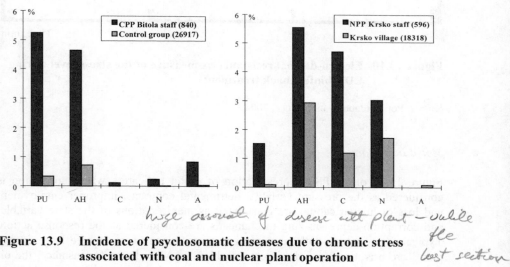

huge amount of disease with plant - while
the least section

Figure 13.9   Incidence of psychosomatic diseases due to chronic stress associated with coal and nuclear plant operation

Source: Pop-Jordanova and Pop-Jordanov, 1996

Note: PU=Peptic ulcer; AH=Arterial hypertension; C=Colitis; N=Neurodermatitis; A=Asthma.

Finally, a possible remedial action can be the application of electro-dermal response biofeedback, i.e. EDR biofeedback modality, as a stress-abating technique. Having the electro-dermal resistance as a measurable indicator of the level of stress, this technique is based upon changing skin conductivity.

A typical result of EDR biofeedback treatment is presented in Figure 13.10. Inversely proportional to the level of stress, the increase in electro-dermal resistance achieved, indicates abatement of occupational stress.

Electro-dermal Resistance [kΩ/10]

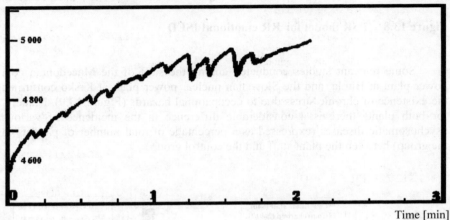

Time [min]

Figure 13.10   Electro-dermal resistance as measure of the stress level during EDR biofeedback treatment

Source: Pop-Jordanova and Zorcec, 2004

Is this about energy + culture ?
Authors of left chapter not speaking to eal other et all

Moral Indicators

Figure 13.11 offers PSR representation of moral indicators where corruption is considered as the pressure variable, detrimental contracts, improper employment and loss of human capital are considered as manifestations of the state variable, and corruption-depressurizing mechanisms are considered as the response action. Typical for countries with economies in transition are cases where large energy plants are privatised in a detrimental fashion. In Macedonia, for instance, the oil company OKTA was sold in a non-transparent and controversial way, and there have been intentions to privatise prematurely the national electric power utility.

overall point

These social deviations have far-reaching consequences for the energy sector, and for the economy in general, so they should not be neglected when analysing the risks associated with various energy technologies.

As a response, governments have manifested more awareness and readiness to introduce various mechanisms to eliminate corruption in the developmental pathways of their societies, but the effects, in particular in countries with economies in transition, are still modest.

privatisation + corruption

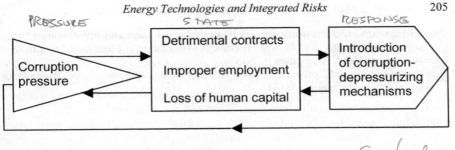

Figure 13.11 PSR model for RR moral ISED

[handwritten: So functional !]

Conclusions *[handwritten: New theme in sust'ble energy research]*

The occupational entropy we have discussed, and the related risks, represent a potential new theme within the social dimension of sustainable energy development.

Until now, the negentropic effects of the shift towards non-material resources and capacities characteristic of the modern economy have been missing from current lists of sustainability indicators. The concept of integrated risks including health and psychological components helps fill this gap. The risk-related mental indicators introduced here, along with the corresponding response actions (biofeedback training and corruption depressurizing) are shown to be particularly important for integrated assessment and management of energy risks.

[handwritten: "corruption depressurizing"! function]

References

Davenport, H.T. and Beck, C.J. (2001), *The Attention Economy*, Harvard Business School Press, Boston, Massachusetts.

IAEA-IEA (2001), 'Indicators for Sustainable Energy Development', *Joint IAEA and IEA Contribution to CSD-9*, New York.

National Research Council (1983), *Risk Assessment in the Federal Government: Managing the Process*, National Academy Press, Washington, DC.

National Research Council (1999), *Our Common Journey*, National Academy Press, Washington, DC.

Pop-Jordanov, J. (2003), 'Indicators for Sustainable Energy Development from a Negentropic Perspective', Original Paper, *Clean Techn. Environ. Policy* 5, pp. 273-278; Published also in (2004): *Technological Choices for Sustainability*, Sikdar, S.K., Glavic, P. and Jain, R. (eds), Springer, pp. 305-316.

Pop-Jordanov, J., Markovska, N., Pop-Jordanova, N. and Simoska, S. (2004), 'Occupational Entropy and Mind Indicators for Sustainable Energy Development', *International Journal of Green Energy*, vol. 1, no 3, pp. 1-9.

Pop-Jordanov, J. and Pop-Jordanova, N. (1990), 'Stress Consequences of Technological emissions', *Developments in Environmental Modeling*, vol. 15, Elsevier, Amsterdam-Oxford-New York-Tokyo, pp. 163-168.

Pop-Jordanova, N. and Markovska, S. (2001), 'Neurofeedback Treatment of Attention Deficit Hyperactivity Disorder', *Proc. XVI Yugosl. Epil. Symp.*, Belgrade.

Pop-Jordanova, N. and Pop-Jordanov, J. (1996), 'Psychosomatic and Substitution Effects: Comparative Health Risks from Electricity Generation', *Electricity, Health and Environment*, IAEA, Vienna, pp. 177-187.

Pop-Jordanova, N. and Zorcec, T. (2004) 'Child Trauma, Attachment and Biofeedback Mitigation', *Contributions, MANU*.

PART 6:
ENERGY AND OPINION

Chapter 14

Measuring and Explaining Environmental Behaviour: The Case of Spain[1]

Juan Díez-Nicolás   *Sociology, Madrid*

Spain ranks in position 76 out of 146 countries on the Environmental Sustainability Index (Esty, Levy, Srebotnjak, and de Sherbinin 2005), it ranks in position 19 out of 22 European Union countries included in the analysis (Luxembourg, Malta and Cyprus are not included), and it ranks in position 23 out of 29 OECD countries. The ESI-2005 index is based on 76 variables, from which only 21 indicators are constructed, which in turn are then grouped into 5 main components (environmental systems, reducing environmental stresses, reducing human vulnerability to environmental stresses, societal and institutional capacity to respond to environmental challenges, and global stewardship). Relatively speaking, Spain seems to score better in terms of reducing human vulnerability to environmental stresses and in societal and institutional capacity to respond to environmental challenges, but it scores worse in environmental systems, global stewardship and in reducing environmental stresses. The ESI report concludes that Spain suffers an overcharge of its ecological systems, though it shows a great capacity to face that problem with success. But it seems quite evident that Spain ranks lower than one would expect when considering measures of economic development (i.e., per capita GNP) or human development (i.e., HDI), where it ranks in position 25 and 20 respectively out of 177 countries (PNUD 2004).

SPAIN ie good at people not at planet

Another international comparative research that measures environmental performance in 17 industrial democracies also shows very poor results for Spain (Scruggs 2003). The research focuses on the measurement and explanation of reducing environmental pollution since the early 1970s. Its results show that Ireland and Spain are the two countries (out of the 17 industrial democracies that are compared) with the lowest reduction in environmental pollution during the period 1970-1995. In this case, however, it must be taken into account that all other

---

[1] The data that provide the basis for discussing the Spanish case were the result of a research grant from the Obra Social Caja Madrid, and were first published in *El Dilema de la Supervivencia (The Survival Dilemma)*, (Díez-Nicolás 2004). The present paper, however, incorporates new ideas, arguments and data that are the result of many fruitful academic discussions with colleagues, including the workshop at International University Bremen.

sixteen countries would rank higher than Spain in terms of GNP per capita and in terms of HDI (except Italy). Scruggs differs from the authors of the ESI report in that he focuses on a selected number of countries (i.e., 17 industrial democracies), in that he focuses only on environmental performance (and more concretely, on the results in reducing environmental pollution), and in that he makes an effort to explain differences in environmental performance through different variables: structural variables (i.e., environmental policies, changes in per capita income, geographic size and density), individual variables (i.e., expressed social concern about environmental protection, post-materialist values), and institutional variables (i.e., economic and political institutions).

One of the most important contributions of Scruggs' research is his insistence on the difference between policies and outcomes, that is, between the intentions expressed by the government towards the environment and the actual results achieved in reducing pollution (performance). But this difference between attitudes and intentions towards the environment and actual environmental behaviour can also be detected at the individual level. In a previous attempt to compare seventeen countries with respect to their orientation towards the environment (Díez-Nicolás 1999) a distinction was made between knowledge, concern, attitudes, intentions to behave, and reported behaviour towards the environment, providing enough evidence to conclude that post-materialist values were positively correlated with knowledge, concern, attitudes, intentions to behave and reported environmentalist behaviour, though it was underlined that reported behaviour on behalf of the environment was very rare in most of the compared countries. On the other hand, it was also found that social position, an indicator that allows to differentiate between the social center (elites) and the social periphery (publics), was also positively related with all the environmental measures that have been mentioned, besides being highly correlated with post-materialist values, thus confirming Galtung's theory about the emergence and change of social attitudes (Galtung 1976). Post-materialist values and social position were also significantly correlated with membership in associations related to environmentalism, a finding that does not contradict Scruggs' finding that membership in environmental associations is not significantly correlated with better performance in reducing pollution in the countries he studied. The two findings actually seem to suggest that too frequently attitudes are taken as proxies of real behaviour, whereas the two often differ considerably.

As a matter of fact, a review of the literature shows that many authors tend to take knowledge about the environment, concern about the environment, attitudes towards the environment, or even intentions to behave towards the environment, as real measures of environmental behaviour. But this inference is questioned when one is confronted with the fact that, though post-materialist values are correlated with membership in environmental associations, membership in associations is not significantly correlated with good environmental performance as measured by reduction in pollution (Scruggs 2003). One hypothesis that seems to arise from these findings is that though attitudes may appear to be positively correlated with behaviour, they may also be a consequence of a desire to adapt to 'political correctness'. There is a large amount of literature on comparative analysis concerning environmental knowledge, concerns, attitudes, mobilization, and intentions or expectations of behaviour (Skrentny 1993, Dunlap 1995, Ellis and Thompson 1997, Gilroy and Shapiro 1986, Rohrschneider 1990, Hofrichter and Reif 1990) but very few references to real behaviour or reported-recalled behaviour. And just as the policies formally adopted or announced by a particular government do not really tell us much about their implementation and results, likewise knowledge, concerns, attitudes and intentions to behave do not tell us much about what individuals' real environmental behaviour will be.

[handwritten margin note: need to be sceptical about both policy claims and individual behaviour wrt environment]

The Theoretical Framework

The most general hypothesis that will be tested in this paper is that attitudes precede behaviour, but they do not necessarily determine it. Attitudes, and indeed opinions (which are overtly expressed attitudes but not real attitudes) may be the result of personal reflection on acquired information about an object, but a person may also acquire them as part of the information itself, without reflecting personally about it (Katz 1953). That may be the reason why in the above-mentioned analysis of 17 countries (Díez-Nicolás 1999) it was found that attitudes (expressed opinions, including intentions to behave) towards the environment were much more frequent and positive than reported behaviour (even though it may be assumed that respondents probably overestimated their positive actions towards the environment). Nevertheless, in spite of a possible adaptation to what seems to be the 'politically correct' orientation, the fact that people feel and express the idea that they should behave properly towards the environment seems to be a result of what Durkheim called 'la contrainte social' (Durkheim 1893), the social pressure, which is not a random event, but a result of other social phenomena.

According to 'social ecosystem' theory (Duncan and Schnore 1959; Duncan 1964; Díez-Nicolás 1982; Hawley 1986), social attitudes are instrumental collective responses that a population develops in order to achieve the best adaptation possible to their environment, under a given state of the arts (technology). Human populations, as all other biotic populations, must interact with their environment and survive through the use of the resources that they find in it. But, contrary to all other biotic populations, human populations always interact with their environment through culture: a material culture (broadly defined as technology) and a non-material culture (broadly defined as social institutions, belief and value systems). Both of these form collective responses. which once established may facilitate or prevent further development, including technological development. The four elements of the ecosystem (i.e., population, environment, social organization and technology), therefore, interact with each other, each one of them being a dependent or an independent variable with respect to the other three, based on the analytical perspective that one adopts. The history of mankind may be very briefly described as a continuous process of change, because any equilibrium is by definition unstable, thus requiring constant new attempts to reach a stable balance among the four elements, a situation that is never attainable. The long-term historical development of human societies has implied growing population,

[handwritten note at bottom: grand theory of "social ecosystem"]

continuous expansion of the environment (mainly due to technological achievements in the fields of transportation and communication), ever-increasing complexity of technology, and ever-changing social institutions (i.e. economic, political, family, spatial, etc.) as well as belief and value systems (i.e. religions, ideologies, social movements, etc.).

If one accepts this very general theoretical framework, the conclusion is that 'culture matters', that attitudes and ideologies are both a result and a pre-requisite of certain other system conditions. The Protestant ethic may well have been a prerequisite for the historical emergence of capitalism (Weber 1905), and the achievement motivation (McClelland 1961) probably was a prerequisite for industrialization and economic development, but both social attitudes were also the result of previous social problematic conditions that required new social responses. Thus, the so called protestant ethic could be interpreted as a collective instrumental response to the need that emerged when capital investment in agriculture was producing very low returns and flourishing trades with the newly discovered worlds were yielding large capital gains that had to be reinvested in new and more productive activities different from agriculture. And the achievement motivation may be interpreted as a collective instrumental response to accelerate the economic recovery after World War Two, stimulating individuals to work more and with higher productivity by attaching more importance to individual merit than to family origins.

Similarly, one could argue that the diffusion of industrialization from the more developed countries (mainly North America, Europe, Japan and Australia) to the rest of the world produced better living conditions for the less developed regions but put also a huge pressure on the environment. This pressure was caused not only by the exponential increase in the intensive use of resources that derived from the exponential growth of population, but also by the fact that industrialization gave mankind, for the first time in its history, the power to destroy all kinds of life on earth. The increasing success of world industrialization at the end of the sixties and at the beginning of the seventies led human societies to become aware of the increasing importance of the environment, not only because of the economic limits of growth (Meadows et al. 1972), but also because of the social limits (Hirsch 1978) and the real threat to life on earth (Toffler 1975). Consequently, concern abut the environment emerged as a collective response to avoid the threats of an intensive use of world resources that resulted from an unforeseen success in achieving world industrialization.

But the emergence of a new reality was not recognized by everybody everywhere at the same time. As 'centre-periphery' theory proposes, new attitudes and values are first developed at the centre of society and then spread towards the social periphery (Galtung 1964, 1976; van der Veer 1976; Díez-Nicolás 1966, 1968, 1995, 1996). Concern about the environment emerged at the end of the sixties in the more developed societies, and within them, among those individuals in higher social positions (the 'social centre' as defined by Galtung), that is, among better informed persons and those with more influence on public opinion. Inasmuch as concern for the environment was at that time a new social issue, it was only natural that it first became adopted by the elites that make the 'social centre' of the more 'central' (developed) societies, as manifested in the growth of international and national new organizations (United Nations 1987, 2003; UNEP 1999) dealing with the environment, as well as in the growth of publications and new lines of research on environmental issues in all domains of science (natural and social). *[handwritten: environ + post materialsm]*

Inglehart's theory of cultural change placed concern for the environment as one of the key indicators of the new post-materialistic orientation that replaced the value system underlying the industrialization process, that is, the new set of self-expression values that characterize post-modern and more developed societies, in contrast to the scarcity or survival values that characterized traditional and pre-industrial societies (Inglehart 1971, 1977, 1990, 1997; Inglehart et al. 2004). The relationship between post-materialist or self-expression values (which include concern for the environment) and environmental knowledge, attitudes and behaviour has been the object of comparative analysis of societies with very different levels of economic and political development (Boltken and Jagodzinsky 1985; van Deth 1983; Duch and Taylor 1993; Gendall et al. 1995; Skrentny 1993; Scruggs 2003).

The theoretical-logical relationship among the three theoretical frames discussed above has not gone unnoticed. In fact, in previous writings it was verified for a number of countries with very different levels of economic development, political organization and cultural values that knowledge about the environment, concern for the environment, preference for protecting the environment over economic development, and intentions to act in favour of the environment, are generally more prevalent in more developed ('central') societies, more prevalent in every society among individuals with higher social positions ('social centre'), and more prevalent in every society among individuals who show a more post-materialist orientation (Díez-Nicolás 1992, 1995, 1999, 2000). But one relationship lacked verification, mainly because measurement instruments used on the surveys from which data were obtained were not appropriate for that purpose: the relationship between attitudes towards the environment and environmental behaviour. This is the main purpose for using Spain as a case study for testing a model that focuses on environmental behaviour as the main dependent variable, and for including attitudes towards the environment, concern and knowledge about the environment, social position, and other relevant variables, as independent explanatory variables. The choice of Spain is especially relevant because, as was mentioned above, it is a country in which environmental policies established by the government do not seem to be implemented, a country that shows a low Environmental Sustainability Index relative to other measures of economic development and ranks among the lowest in terms of reducing environmental pollution (Esty et al. 2005; Scruggs 2003).

[handwritten: long-winded introduction to a focus on envir behaviour in Spain]

The Measurement of Environmental Behaviour

The theoretical model used to explain environmental behaviour in Spain consists of a path analysis model in which the dependent variable is a compound index of environmental behaviour, and the six independent variables are all the product of a

[handwritten: dependent + independent / explanatory variables]

combination of items included in a questionnaire applied to a representative sample of 1,224 residents in Spain, 18 years old and over, through face-to-face interviews in their homes. The sample design starts with the proportional distribution of interviews among the 17 regions according to their population and to community size within each region. Municipalities with more than 500,000 inhabitants are of compulsory selection; the rest come out of a random draw. Once the number of interviews has been established (by size of municipality and region), municipalities are randomly extracted through a computerized system. Electoral sections, generally around 155, are also randomly selected within each municipality. A random route system is applied for household selection within each electoral section. Age and sex quotas within each random route (established for each electoral section on the basis of its census distribution by size of community within each region) are used for selecting the respondent within each household.

The six independent variables, ordered from the most antecedent variable to the last in the path analysis model are the following: social position, environmental information, knowledge about the environment, post-materialism, environmental orientation, and confidence in civil society.

According to the theoretical framework presented above, one would expect to find a positive relationship between social position and post-materialism, on the one hand, and good practices of environmental behaviour on the other hand. These are the two major hypotheses to be tested in the model. But some other hypotheses are also derived from theory. Thus, according to 'centre-periphery' theory, the social centre is more informed and has more knowledge and opinions than the social periphery about any issue, and therefore one would expect to find positive relationships between social position and information on the environment, between social position and knowledge about the environment, between social position and post-materialism (based on the assumption that the social centre internalizes new values earlier than the social periphery), and between social position and an environmental orientation (attitudes more favourable to protecting the environment than to economic development). There is no theoretical reason to expect, however, any particular relationship, positive or negative, between social position and confidence in civil society.

pompous

Regarding information on the environment, and always according to the theoretical assumptions presented above, one would expect to find that individuals who are more informed about the environment should have more knowledge about it, should be more post-materialist oriented, should be more favourable to protecting the environment, and should also have better practices towards the environment. Again, however, there is no reason to expect any particular relationship between information on the environment and confidence in civil society, though one might expect that individuals who are better informed about the environment are also more informed about other issues, due to a higher level of education, which would lead them to trust civil society more than public administrations (Putnam 1993). Following similar arguments, knowledge about the environment should be positively related to post-materialism, to favourable attitudes concerning the protection of the environment, and to good practices of environmental behaviour, but one should not expect any kind of relationship with confidence in civil society. Finally, attitudes

why presume?

blah blah

towards the environment and confidence in civil society should be positively related to good practices of environmental behaviour.

The construction of indexes to measure each variable in the model has followed the following steps. First, the social position index is based on seven socio-demographic variables. The index of social position has been constructed through an adaptation of Galtung's index, and modifying previous adaptations to Spain of that index made by the author (Díez-Nicolás 1968), avoiding dichotomization of variables and giving different (rather than equal) weights to the component variables. The values attached to categories in each variable are the following. Sex (male = 1; female = 0). Age (<18 and >75=0; 18-25 and 65-74 = 1; 26-35 and 55-64 = 2; and 36-54 = 3). Educational level (less than primary and missing = 0; primary, elementary, secondary first cycle, vocational = 1; secondary second cycle, pre-university = 2; university degree = 3). Monthly income (<450 € = 0; 451-900 € = 1; 901-1,650 € = 2; >1,650 € = 3). Size of habitat (<10,000 inhabitants = 0; 10,000-50,000 = 1; 50,000-250,000 = 2; 250,000 plus Madrid and Barcelona = 3). Occupational status (no occupation plus missing = 0; non qualified = 1; qualified and middle status occupations = 2; high status occupations = 3). Economic sector (no occupation plus missing = 0; primary, extractive sector = 1; secondary, industrial sector = 2; tertiary, service sector = 3). Centrality (regions with low per capita income [Castilla-La Mancha, Galicia, Andalucía, Extremadura] = 0; regions with middle per capita income [La Rioja, Aragón, Cantabria, Valencia, Castilla-León, Canarias, Asturias, Murcia] = 1; regions with high per capita income[Madrid, Navarra, País Vasco, Baleares, Cataluña] = 2). The social position index could therefore vary between 0 and 27 points. The correlation coefficient between the index of social position and the more common index of socio-economic status is r=.50, but the former has shown greater predictive value than SES (Díez-Nicolás 1992, 2004).

Social position is positively and significantly correlated at .01 level with the eight component socio-demographic variables, as expected, but the correlation coefficients are especially high with occupation, education, economic sector and income, and lower with sex and age, as was intended when deciding to give more weight to occupation, income and education. As to the distribution of the index, it shows a bell-shaped curve with about 10 percent of respondents in high social positions (21 points or more), but only 3 percent in what Galtung would call 'the decision-making nucleus' (24 points or more). At the other end of the scale, about a quarter of the sample qualifies as 'social periphery' (10 points or less), and about 5 percent could even be considered 'extreme social periphery' (5 points or less).

Information about the environment was measured through the number of sources that individuals said they used to obtain information on the subject. In fact, three measures of exposure to information are considered: one is a 'general index of exposure of information' which takes into account newspaper readership, listening to information programmes from general broadcasts, watching TV news programmes; a second index is based on the respondents' evaluation as to how well informed about environmental issues they feel; and the third index is based on the number of sources that respondents said they used to obtain environmental information. It was found that 15 percent of the sample shows a high index of exposure to general information (every day they read one newspaper, listen to a radio news programme and watch a TV news programme), and 29 percent answers they feel very or rather well informed about environmental issues. However, 29 percent of respondents admits that they do not use any of the thirteen sources of information on environmental issues that were mentioned to them, and only 3 percent that they use five or more of those sources to obtain information on the environment. The thirteen sources of information on environmental issues that were presented to respondents were: newspapers, radio, TV, ecological associations, other scientific associations, internet, studies or professional training, public lectures or courses, professional activity, voluntary work, friends, specialized magazines, and other sources. A main component analysis showed four different factors: one that included the three media sources, a second one that included the two sources about associations, a third one that included the three sources on study and profession, and a fourth one that included only voluntary work. The other three sources (friends, specialized magazines and others) did not fit into any of the four factors or any other factor. TV was undoubtedly the most cited source of information on the environment. A regression model in which exposure to environmental information was the dependent variable and social position, post-materialism and general exposure to information were included as independent variables explained 17 percent of the variance, and though the three variables showed significant standardized regression coefficients, social position seemed to contribute more than the other two to that explanation. Besides, the three indicators of information on the environment are significantly correlated: general information and self-evaluation (r=.19), general information and exposure to information on the environment (r=.25), and self-evaluation and exposure to information on the environment (r=.61).

Environmental culture has been measured through eight items, but a principal component analysis showed that there are two components, one that measures scientific knowledge about the environment (five items), and another one that measures concern about the environment (three items). Statistical analysis demonstrated that only scientific knowledge about the environment was really relevant, though the correlation between the two indicators was r=.47. Scientific knowledge has been measured by giving the correct answers to four statements on the environment: 'If someone is exposed to a certain amount or radioactivity, no matter how small, he/she will certainly die'; 'all pesticides and chemicals used on food crops cause cancer in human beings'; 'some radioactive residues produced by nuclear plants will remain dangerous for thousands of years'; 'every time that coal or oil are used the [green house] effect is worsened'; and 'cellular phone antennas are dangerous for the health of individuals'. Since for each item the respondent could answer 'totally true, probably true, probably false, or totally false', the resulting scale could vary from five totally correct answers (20 points) to five totally incorrect answers (0 points).

Exposure to environmental information shows a greater relationship with scientific knowledge about the environment than with concern about the environment. On the other hand, the items that measure knowledge had all been tested successfully in many other surveys, while the items that measure concern

Some paper repeat again and again,
often are very reflective + discursive   [indigestible detail!]

Measuring and Explaining Environmental Behaviour: The Case of Spain   217

were new and had not been tested before. And, finally, the items that measure concern seem to produce answers very much in line with 'political correctness'. For all these reasons, it was decided to measure this variable only through the scientific knowledge items. It must be underlined that only around 15 percent of the respondents seems to be really knowledgeable about the environment (obtaining 16 points or more), while about the same proportion seems to have a very low knowledge about the environment (obtaining 9 points or less). A regression model to explain knowledge about the environment using social position, post-materialism and exposure to information on environment explains 19 percent of the total variance, and though the three predictors contribute significantly to that explanation, exposure to information seems to contribute less than the other two variables because of its high intercorrelations with them.

Post-materialism has been measured using Inglehart's scale of twelve items. The twelve items were divided into two groups, a first group of four items, two measuring materialist values ('maintaining order in the nation' and 'fighting rising prices') and two measuring post-materialist values ('giving people more say in important government decisions' and 'protecting freedom of speech'), and a second group of eight items, four measuring materialist values ('a high level of economic growth', 'making sure this country has strong defence forces', 'a stable economy', and 'the fight against crime') and four measuring post-materialist values ('seeing that people have more say about how things are done at their jobs and in their communities', 'trying to make our cities and countryside more beautiful', 'progress toward a less impersonal and more humane society', and 'progress toward a society in which ideas count more than money'). Since respondents could choose two items from the first group of four items, and three from the second group of eight items, they could select in total a maximum of 5 and a minimum of 0 post-materialist items. This is an extensively tested scale, regardless of whether one uses only a scale of four items, a scale of four and another one of eight items, or three scales of four items each. Social position and post-materialism are certainly positive and significantly correlated ($r=.16$), as expected, but it must be underlined that their relationship is far from perfect, a finding that supports the decision to include the two indexes as separate independent variables to explain good practices of environmental behaviour. A regression model to explain post-materialism through social position, exposure to environmental information and scientific knowledge on the environment explains 10 percent of the variance, but social position does not add significantly to that explanation in the presence of the other two predictors, information and knowledge on the environment, which contribute more or less the same.

Pro-environmental orientation has been measured through nine items, some of them more favourable to economic development and others more favourable to protecting the environment. To construct the index, a principal component analysis was made with the nine items that measure preferences towards the environment or economic development, and only one factor was extracted, so that they scaled themselves with the most pro-environment at one end and the most pro-development at the other end. Then, the two most pro-environment items ('to protect the environment it is necessary to reduce our consumption and standard of living'; 'the protection of the environment requires more solidarity with the less developed countries'), and the two more pro-development items ('people have the right to use all the artefacts that technology provides, even if when using them we unintentionally deteriorate the environment'; 'it is right to use animals in medical experiments if it helps to save human lives') were selected to construct an index. Since respondents had to agree or disagree with each item on a four point scale, the index varies between 4 (completely disagree with the two pro-environment items and completely agree with the two pro-development items) and 16 (exactly the reverse). The nine items were scaled on a bipolar axis through a principal component analysis with only one extracted factor. The distribution of respondents on a 4 to 16 point scale was again a bell-shaped curve skewed towards the environmentalism pole, with almost 10 percent of respondents on the three more pro-environmental positions, and only less than 1 percent on the three more pro-development positions. The regression model calculated to explain the pro-environmental orientation through the previous four variables (social position, exposure to environmental information, knowledge about the environment and post-materialism) explains only 6 percent of the total variance, and only social position and post-materialism show significant contributions to explaining that variance.

Finally, confidence in civil society has been measured on the basis of four questions that attempted to asses the degree of confidence that respondents had in different civil institutions (educational, mass media and business and industrial firms) regarding the protection of the environment. The index of 'confidence in civil society' was built on the basis of four questions: confidence on the school education that the respondent received concerning the protection and maintenance of the environment; opinion on whether the information on the environment provided by the press, broadcasting stations, and TV channels is sufficient or insufficient; opinion on whether or not business and industrial firms take into account environmental criteria in their processes of production and manufacturing; and opinion on whether or not business and industrial firms give at present more or less information on the ingredients and components of their products or over their impact on the environment.

The scale varies between 7 and 29 points, but one third of respondents did not give an answer to any of the four questions. A regression model constructed to explain confidence in civil society shows that only 4 percent of its variance is explained by the five previous variables, though only exposure to environmental information and attitudes towards the environment contribute significantly to that explanation. It must be underlined, however, that ideology (measured on a self-positioning scale of seven points) is very significantly related to confidence in civil society ($r=.20$), suggesting that individuals who place themselves on the right tend to have greater confidence in civil society than those who place themselves on the left (a finding that is coherent with the complementary finding that individuals who place themselves on the left tend to rely more on public administration than on civil society).

But the major goals of this paper were to measure environmental behaviour and to explain why some individuals show better practices of behaviour towards the environment. Several approaches have been tested to measure behaviour

Measurny behaviour

Measuring and Explaining Environmental Behaviour: The Case of Spain   219

towards the environment because behaviour as such cannot be measured through surveys, but only reported behaviour or intentions to behave. First, respondents were asked for the frequency with which they practiced a total of twenty-three common activities that may have an impact on the environment. The list of activities was: driving a car, driving a motorcycle, driving a work vehicle (bus, truck, tractor, etc.), double parking, smoking at home or in open spaces, throwing trash in the streets, separating garbage in different bags, using sprays, throwing cigarettes or trash to the floor in bars or cafes, depositing newspapers and other papers in containers, leaving garbage bags and other rubbish (bottles, cans, etc.) in the countryside or in beaches after a picnic, throwing cigarettes or trash out of the car's window, lighting a fire in the fields or woods, smoking at work, bars or restaurants, or in any other indoor space, opening a tap and letting the water run unnecessarily, throwing away batteries with the regular garbage, leaving lights on in rooms where there is nobody, buying products using the least possible amount of wrapping, throwing bread or other food products into the garbage bag because of expired date, using public transportation for daily activities instead of private car, buying recycled paper or products, depositing bottles in the appropriate containers, burying cigarettes butts in the sand at the beach, and other activities. The most frequently practised activities are driving a car (86 percent), smoking at home or in open spaces, throwing papers and other rubbish in the street, smoking at work, in bars, restaurants and other indoors public places, and throwing cigarettes and/or trash to the floor in bars and cafes (between 65 percent and 61 percent). The twenty-three activities were then classified as good or bad practices towards the environment (seven of them were classified as good and sixteen as bad practices), and an index showing the difference between good and bad practices for each individual was constructed. The classification of these practices as good or bad was confirmed through a principal components analysis extracting only one factor. For each individual only those activities that were practised sometimes or usually were taken into account. The index could vary between -16 and +7, but 100 were added to the result to avoid negative values, so that the scale could vary between 84 and 107 points.

A second index was calculated taking into account the frequency of practising each activity (i.e., giving different weights according to frequency of practice) and the degree of damage to the environment that individuals attributed to each activity. For the seven 'good' practices 3 points were given if they were practiced usually, 2 points if sometimes, and 0 points if never. 'Bad' practices were separated into two groups, one including the eight that were considered as more damaging to the environment by respondents, and a second group including the other eight activities. For the activities in the first group 0 points were given to those who said they practiced them usually, 1 point if practiced sometimes, and 3 points if never practiced. For the activities in the second group 0 points were given if practised usually, 1 point if practised sometimes, and 2 points if never practised. This index could vary between 0 and 61 points.

A third index based on the same data was built, taking into account only the seven 'good practices'. This index was very simple, as it only took account of whether or not each of the seven 'good' activities was ever practised by the

relying on self-disclosure *

respondent, so that the values could vary between 0 and 7. Only 9 percent of respondents said they had practised all seven good practices, while 5 percent had practised none.

Another approach to measuring good behaviour towards the environment referred to reported changes in a consumer's behaviour. The questions, in this case, asked whether or not respondents had changed their habits of water, gas or electricity consumption, and their buying habits, in order to save energy or to take into account environmental protection criteria. In every case, if the answer was positive (i.e., R had changed habits to save energy or to protect the environment) 3 points were given; two points were given if the answer was 'no, because I use only what is necessary' or 'no, because I changed my habits before'; and one point was given if the answer was 'no, I did not change them because I don't care about those things'. The index could vary between 4 and 12 points, but 17 percent of respondents did not answer these questions, and only 9 percent had changed the four habits in order to save energy or protect the environment.

Consumer behaviour has also been measured through some other questions that have been used to elaborate another index. In this case five different consumption habits were taken into account: looking at expiration date of food products, buying household appliances of low energy consumption, buying house cleaning products that are not aggressive to the environment, buying recycled products, and buying fruit and vegetables non-exposed to pesticides or chemical products. Since the frequency for each one of the five consumption habits was available, 2 points were given to those habits that were followed always or almost always, 1 point if followed sometimes, and 0 points if followed never or almost never. The index could vary between 0 and 10 points, and 15 percent of respondents did not answer the questions, but while 8 percent obtained 3 points or less (on a scale 0 to 14), 14 percent obtained 7 or more points.

One affirmative action index has been constructed through four items that asked about participation in political activities in favour of the environment: membership in some group or association engaged in protecting the environment, signing some collective letter within the last 5 years for some environmental cause, giving money to some environmental group, or participating in some protest or public demonstration on some environmental issue. Each individual received one point for each activity in which he had ever engaged, so that the index could vary between 0 and 4 points. It must be underlined than more than 80 percent of respondents had never done any of the four activities, while only 13 persons had done all four. This finding by itself is a clear demonstration of the great gap between attitudes and behaviour, and why it is so important to measure behaviour, even at the risk of overestimating good behaviour due to the fact that it is necessary to rely on the respondent's answers.

The last index was intended to measure 'disposition' to behave in favour of the environment. Only two items were used to construct this index: one asked if R would be in favour or against paying more taxes or accepting a decrease in their present standard of living in order to protect the environment. The scale for each item was one of five categories: 'very much in favour', 'somewhat in favour', 'nor in favour neither against', 'somewhat against' and 'very much against', giving from 5 to 1, and 0 for no answer. The scale could vary from 0 to 10. The correlation coefficient between the two items was r=.60 and statistically significant.

It is interesting to note that, in contrast with the little action measured by the answers regarding real behaviour (which probably were somewhat exaggerated), when the questions refer to 'intentions' to behave individuals seem to be much more ready to act. In fact, 26 percent of the respondents say that they would accept paying more taxes if they were applied to protect the environment, and 36 percent would accept lowering their standard of living in order to protect the environment. The contrast between future expectations and past actions is remarkable, and certainly warns against using 'intentions' to behave as good predictors.

To summarize, seven indexes of environmental behaviour have been constructed. Using regression models with the model independent variables and each index as the dependent variable, the corrected explained variance varies between 14 percent and 26 percent for six indexes, but only 2 percent with respect to the index measuring change of consumption habits. In most models social position and post-materialism are the variables with the highest standardized regression coefficients. But it must be admitted that using seven indexes as the dependent variable does not clarify the measurement of environmental behaviour. Therefore, and in view of the knowledge gained through the very detailed analysis that was performed, a summary index for measuring environmental behaviour was constructed. This index has been based on all the items that imply good practices towards the environment, giving one point for each. The items are the following: separating garbage in different bags, depositing newspapers and other papers in appropriate containers, buying products with the least possible amount of wrapping, placing in the trash bread or other food products whose expiration date has passed, using public transportation for daily activities instead of private car, buying recycled paper or other recycled products, and depositing bottles in appropriate containers. One point was also given for reducing the use of water, gas and electricity and for modifying consumption habits to care for the environment. One point was given for doing always or almost always the following: looking at the expiration date of food products, buying ecological food products grown naturally, buying house cleaning products that are not aggressive towards the environment, buying recycled products, buying products with the 'ecological label', buying household appliances with low energy consumption, buying fruits and vegetables grown without pesticides or chemical products, and giving up driving the car for environmental reasons. One point was also given for being a member of a group or association engaged in protecting the environment, having signed some collective letter about some environmental issue, having given money to some environmental group, or having participated in some protest group or public demonstration for some environmental cause. The total number of good practices is twenty-three, so that the index could vary between 0 and 23.

The global 'index of good environmental practices' can vary between 0 and 23 points, and it may be seen that only less than 5 percent of respondents obtain 13 or more points, while 55 percent obtain 5 or less points. It is quite evident that, on the basis of such a variety of possible good practices towards the environment, the great majority of Spaniards obtain a very low score. And it must be emphasized that bad practices (which, as has been demonstrated, are quite frequent) have not been included in this index, and neither have intentions to behave, since the answers seem to be rather exaggerated.

To test the validity and reliability of this global 'index of good environmental practices', a correlation matrix of the previous seven indexes and the new global index has been calculated. The main conclusion that can be derived from this correlation matrix is that the global 'index of good environmental practices' shows the strongest correlation coefficients with all other indexes, a finding that seems to guarantee its utility to measure good environmental behaviour, and that consequently fulfils one of the main goals of this chapter.

Table 14.1   Correlation coefficients (Pearson's r) among the different indexes of environmental behaviour*

		(1)	(2)	(3)	(4)	(5)	(6)	(7)	(8)
(1)	Global index of good environmental practices	-							
(2)	Difference between positive and negative behaviours	.26	-						
(3)	Behaviour scaling and frequency of practice	.35	.97	-					
(4)	Favourable behaviours (positive, good practices only)	.44	.44	.51	-				
(5)	Change in consumption habits	.53	.12	.14	.14	-			
(6)	Ecological behaviours	.55	.10	.16	.26	.17	-		
(7)	Participation in activities of affirmative action	.47	(.05)	(.06)	.21	.11	.25	-	
(8)	Intentions to behave in favour of the environment	.28	(.07)	.09	.20	.17	.18	.22	-

* All coefficients are significant at .01 level except those between brackets.

The Explanation of Environmental Behaviour

A path analysis model has been constructed to explain behaviour towards the environment. The dependent variable is the 'global index of good environmental practices' as defined above. The most antecedent independent variable is social position, and the intervening variables are exposure to environmental information, scientific knowledge about the environment, post-materialist values, pro-environment orientation, and confidence in civil society.

Table 14.2   **Correlation coefficients (Pearson's r) among the different independent variables and the global index of good environmental practices***

	(1)	(2)	(3)	(4)	(5)	(6)	(7)
(1) Global Index of good environmental practices	-						
(2) Social position	.26	-					
(3) Post-materialism	.26	.16	-				
(4) Exposure to information on the environment	.29	.34	.24	-			
(5) Scientific knowledge about the environment	.22	.26	.26	.25	-		
(6) Environmental orientation	.14	.11	.24	(.04)	(.03)	-	
(7) Confidence on civil society	.14	(-.00)	(-.04)	.12	(.03)	-.16	-

* All coefficients are significant at .01 level except those between brackets.

All independent variables are positively and significantly correlated with the global index of good environmental practices, and most of the correlations among the independent variables are also positive and significant, but two of the variables (environmental orientation and confidence on civil society) do not show strong relationships with the other variables, including the global index of good environmental practices. Most interesting is that exposure to information on the environment and scientific knowledge concerning the environment are not significantly correlated with attitudes toward the environment. However, post-materialism shows the strongest relationship with it. Besides, confidence in civil society is not related to social position or to post-materialist values, neither to knowledge about the environment, but individuals who are more exposed to environmental information have greater confidence in civil society. And those who are more in favour of economic development than of protecting the environment have less confidence in civil society. The reason, as explained above, is that individuals who politically place themselves on the left are more favourable to the public sector than to civil society, but more favourable to the environment than to economic development, while those who place themselves on the right trust civil society more than they trust the public sector, and they are more favourable to economic development than to protecting the environment. Ideology, however, was not included into the path analysis model as an intervening explanatory variable because it is not significantly related at all to behaviour towards the environment.

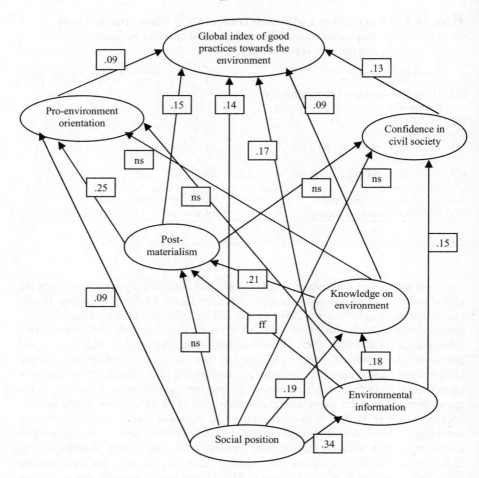

Figure 14.1   Standardized direct effects of each one of the explanatory variables in the model on all other variables

The standardized direct effects of each variable on all others, following the path established in the model, confirms some hypotheses that have been repeatedly verified by research. Thus, the 'social centre' is more informed and has more knowledge about everything (in this case about the environment) than the 'social periphery'. The social centre shows better environmental behaviour than the social periphery, thus confirming also that the social centre internalizes new values (good practices towards the environment) earlier than the social periphery (Galtung 1964; Díez-Nicolás 1968). But social position has no direct effect on confidence in civil society nor on post-materialist values (an apparently surprising finding that will be explained later, because these values are new, and therefore should be internalized earlier by the social centre than by the social periphery), and a very weak though significant direct effect on attitudes in favour of the environment. Post-materialism, which is the other major explanatory variable according to the theoretical framework, shows significant direct effects on preference for the environment over economic development, confirming Inglehart's assumptions and findings (Inglehart 1977, 1990), and also on environmental behaviour, suggesting that values have an impact on behaviour but no direct effects on confidence on civil society. The model also confirms that exposure to information and knowledge about the environment have no direct effects on preferences for protecting the environment over economic development, suggesting that attitudes may be a consequence of adaptation to what seems 'politically correct'.

Policy implications?

Table 14.3   Effects of explanatory variables on good practices towards the environment*

	Non-standardized effects		
	Direct +	Indirect =	Total
Social position	.11	.08	.19
Exposure to environmental information	.45	.18	.63
Scientific knowledge on the environment	.09	.03	.13
Post-materialist values	.45	.04	.50
Confidence in civil society	.17	--	.17
Attitudes favourable to environment	.17	--	.17
	Standardized effects		
	Direct +	Indirect =	Total
Social position	.14	.11	.26
Exposure to environmental information	.17	.07	.23
Scientific knowledge on the environment	.09	.03	.12
Post-materialist values	.15	.01	.16
Confidence in civil society	.13	.00	.13
Attitudes favourable to environment	.09	.00	.09

* All coefficients are significant at .05 level.

Furthermore, the six explanatory variables in the model have a direct and statistically significant relationship to the dependent variable, though knowledge about the environment and preference for the environment over economic development have weaker though significant relationships to it. This means that individuals who belong to the social centre, who have a post-materialist value orientation, who are more exposed to information on the environment, who have more knowledge about the environment, who attach more importance to protecting the environment than to economic development, and who show more confidence on civil society to protect the environment, tend to behave in a more favourable

Long-winded summary

manner towards the environment than their respective counterparts. The model explains 18 percent of the total variance on the good practices towards the environment, an important proportion when one considers the complexity of the model and, in particular, the complexity of the dependent variable itself.

Table 14.4   **Standardized effects of explanatory variables among themselves***

	Social Position	Exposure to information	Knowledge on environment	Post-materialism	Confidence in civil society	Attitudes towards environment
	Direct effects					
Exposure to information	.34*					
Scientific knowledge	.19*	.18*				
Post-materialism	.05	.18*	.21*			
Confidence in civil society	-.04	.15*		-.06		
Environmental attitudes	.09*	-.03	-.02	.25*		
	Indirect effects					
Exposure to information	-					
Scientific knowledge	.06					
Post-materialism	.11	.04				
Confidence in civil society	.04	-.01	-.01			
Environmental attitudes	.02	.05	.05			
	Total effects					
Exposure to information	.34					
Scientific knowledge	.26	.18				
Post-materialism	.16	.21	.21			
Confidence in civil society	-.00	.13	-.01	-.06		
Environmental attitudes	.12	.02	.03	.25		

* These coefficients are significant at .05 level.

Undoubtedly, one of the most important findings is that attitudes towards the environment do not seem to be a consequence of being informed or having knowledge about the environment. But this is not the only realm of social life where attitudes are accepted without the necessary reflection. Mass media have made possible a massive transmission of values, attitudes and opinions that are accepted without critical reflection by large sectors of the population and which do not respond to deep convictions, but to a predisposition and desire to adapt to what is taken as the majority opinion or as the more socially acceptable.

Another important finding is the lack of significant and direct effects of social position on post-materialism. The significance level required for an error of ± 5 percent is > 1.96. Standardized direct effects of social position on post-materialism, as well as that of social position on confidence in civil society, are very close to this level but do not achieve it.

But it must be noted that though the direct effects of social position on post-materialism are small (but positive), the indirect effects (through exposure to information and scientific knowledge about the environment) are very strong and positive. This means that not all individuals in the social centre adopt post-materialist values, but only those who, in addition, receive more information and have more knowledge about the environment. It is also important to underline that the greater explanatory power of social position with respect to post-materialist values is not new (Díez-Nicolás 1999), and this seems to result from the explanation advanced when discussing the theoretical framework, and more concretely when discussing the social-ecosystem and the centre-periphery theories. The confirmation of a primacy of social position over post-materialist values in explaining behaviour towards the environment should not be interpreted as a rejection of Inglehart's theory, but only as a specification of it that results from the assumption that attitudes (and more so behaviours) towards the environment are changing not only because of the modernization and post-modernization processes analyzed by Inglehart, but also because the social centre has become conscious of the real threat created by mankind to the survival of life on earth. For this same reason, it seems plausible that attitudes favourable to protecting the environment are being transferred from the social centre to the social periphery with greater intensity and speed than behaviours, since the people in the social periphery try to adapt their opinions (probably not as much as their real attitudes) to what they accept as 'politically correct', but without really internalizing these attitudes, and therefore, without this adaptation implying an effective translation of attitudes into behaviours towards the environment. It cannot be overlooked that, while 32 percent of respondents declare their readiness to lower their lifestyle significantly in order to protect the environment better, only 5 percent declare having contributed money to some ecological or environmental organization or group. The contradiction between attitudes and behaviours that these results show is not new, but common to other surveys conducted in Spain and other countries (Díez-Nicolás 1999). The contradiction does not result necessarily from a deliberate intention to lie, but from a process that is taking place in many societies, whereby attitudes are being transferred earlier and quicker than behaviours from the social centre to the social periphery, a process which is normal with respect to many other social changes. In other words, the majority of Spaniards, and probably of other nationals, really believe that they 'should' give priority to protecting the environment over economic development, but their real behaviours and value orientations continue to give greater priority to economic development.

Not attempt to draw at any political / policy applications - disappointing.

References

Boltken, F. and W. Jagodzinsky (1985), 'In an environment of insecurity: postmaterialism in the European Community, 1970 to 1980', *Comparative Political Studies*, 17.

van Deth, J.W. (1983), 'The persistence of materialist and postmaterialist value orientations', *European Journal of Political Science*, 9.

Díez Nicolás, J. (1966b), 'Posición social y opinión pública' (Social position and public opinion), *Anales de Sociología*, 2: 63-75.

Díez Nicolás, J. (1968), 'Social position and attitudes towards domestic issues in Spain', *Polls*, III, 2: 1-15.

Díez Nicolás, J. (1982), 'Ecología humana y ecosistema social' (Human ecology and the social ecosystem), in CEOTMA, *Sociología y Medio Ambiente* (Sociology and Environment). MOPU, Madrid.

Díez Nicolás, J. (1992), 'Posición social, información y postmaterialismo', *Revista Española de Investigaciones Sociológicas*, 57: 21-35. (Trad. al ingles (1996): 'Social position, information and postmaterialism', *REIS*, English edition: 153-165.)

Díez Nicolás, J. (1995), 'Postmaterialism and the social ecosystem', in Beat and Beatrix Sitter Liver (eds.), *Culture Within Nature*. UNESCO, Paris.

Díez Nicolás, J. (1999), 'Industrialization and concern for the environment', in N. Tos, P.Ph. Moler y B. Malnar (eds.), *Modern Society and Values*. FSS y Mannhemim, ZUMA, Ljubljana.

Díez Nicolás, J. (2000), 'La Escala de postmaterialismo como medida del cambio de valores en las sociedades contemporáneas' (The scale of postmaterialism as a measure of value change in contemporary societies), in F. Andrés Orizo and J. Elzo, *España 2000, entre el Localismo y la Globalidad. La Encuesta Europea de Valores en su Tercera Aplicación, 1981-1999.* (Spain 2000, between localism and globality. The European Values Survey in its Third Application, 1981-1999.) Editorial Santa María, Madrid.

Díez Nicolás, J. (2004), *El Dilema de la Supervivencia (The Dilemma of Survival)*. Obra Social Caja Madrid, Madrid.

Duch, R.M. and M.A. Taylor (1993), 'Postmaterialism and the economic condition', *American Journal of Political Science*, 37.

Duncan, O.D. (1964), 'Social organization and the ecosystem', in: R.E.L. Faris (ed.), *Handbook of Modern Sociology*. Rand Mc Nally and Co, Chicago.

Duncan, O.D. and Schnore, F. (1959), 'Cultural, behavioral and ecological perspectives in the study of social organization', *The American Journal of Sociology*, LXV: 132-153.

Dunlap, R. (1995), 'Public Opinion and Environmental Policy', in J. Lester (ed.), *Environmental Politics and Policies*, 63-113. Durham, Duke University Press, NC.

Durkheim, E. (1893), *De la Division du Travail Social*. Alcan, Paris.

Ellis, R.J. and F. Thompson (1997), 'Culture and the Environment in the Pacific Northwest', *American Political Science Review*, 91: 885-98.

Esty, D.C., M. Levy, T. Srebotnjak and A. de Sherbinin (2005), *2005 Environmental Sustainability Index: Benchmarking National Environmental Stewardship*. Yale Center for Environmental Law and Policy, New Haven.

Galtung, J. (1964), 'Foreign policy opinion as a function of social position', *Journal of Peace Research*, 34: 206-231.

Galtung, J. (1976), 'Social position and the image of the future', in H. Ornauer and others (eds.), *Images of the World in the Year 2000*. Mouton, Paris.

Gendall, P., Smith, T.W. and Russell, D. (1995), 'Knowledge of scientific and environmental facts: A comparison of six countries', *Marketing Bulletin*, 6: 65-74.

Gilroy, J. and R. Shapiro (1986), 'The Polls: Environmental Protection', *Public Opinion Quarterly*, 50: 270-79.

Hawley, A.H. (1986), *Human Ecology. A Theoretical Essay*. Chicago: The University of Chicago Press. Traducción castellana (1991): *Teoría de la Ecología Humana*.Tecnos, Madrid.

Hirsch, F. (1978), *Social Limits to Growth*. Harvard University Press, Cambridge.

Hofrichter, J. and K. Reif (1990), 'Evolution of Environmental Attitudes in the European Community', *Scandinavian Political Studies*, 13 (2): 119-46.

Inglehart, R. (1971), 'The silent revolution in Europe: intergenerational change in post-industrial societies', *American Political Science Review*, 65.

Inglehart, R. (1977), *The Silent Revolution*. Princeton University Press, Princeton.

Inglehart, R. (1990), *Culture Shift in Advanced Industrial Society*. Princeton University Press, Princeton.

Inglehart, R. (1997), *Modernization and Postmodernization*. Princeton University Press, Princeton.

Inglehart, R., M. Basañez, J. Díez Medrano, L. Halman and R. Luijkx (2004), *Human Beliefs and Values*. Siglo XXI, Mexico.

Katz, D. (1953), 'Three criteria: knowledge, conviction and significance', in B. Berelson and M. Janowitz, *Public Opinion and Communication*. The Free Press, Glencoe, Ill.

McClelland, D.C. (1961), *The Achieving Society*, D. van Nostrand Co, New Jersey.

Meadows, et al. (1972), *The Limits to Growth*. Universe Books, New York.

PNUD (Programa de Naciones Unidas para el Desarrollo) (2004), *Informe sobre Desarrollo Humano 2004*. Ediciones Mundi Prensa, Madrid.

Putnam, R.D. (1993), *Making Democracy Work: Civic Traditions in Modern Italy*. Princeton University Press, Princeton.

Rohrschneider, R. (1990), 'The Roots of Public Opinion toward New Social Movement: An Empirical Test of Competing Explanations', *American Journal of Political Science* 34: 1-30.

Scruggs, L. (2003), *Sustaining Abundance*. Cambridge University Press, Cambridge.

Skrentny, J.D. (1993), 'Concern for the environment: A cross national perspective', *International Journal of Public Opinion Research*, 5: 335-354.

Toffler, A. (1975), *The Ecospasm Report*. Bantam Books, New York.

United Nations (1987), *Our Commnon Future*. UN Commission on Environrnent and Development. United Nations, New York.

United Nations (2003), *The Road from Johannesburg: World Summit on Sustainable Development*. United Nations, New York.

UNEP (United Nations Environment Program) (1999), *Global Environment Outlook*. Earthscan Publications, London.

van der Veer, K. (1976), 'Social position, dogmatism and social participation as independent variables', in: H. Ornauer et al. (eds.), *Images of the World in the Year 2000*, Mouton, Paris.

Weber, M. (1905/1958/1976), *The Protestant Ethic and the Spirit of Capitalism*. tr. Talcott Parsons, intro. A. Giddens.

Afterword

Brendan Dooley

In December 2004, at the time of writing, a natural catastrophe of epic proportions had just struck Southeast Asia killing well over a hundred and eighty thousand people and leaving many more injured and homeless. The social sciences are not a rescue team; but our institutions, our democracies, our fellow citizens, have given us the responsibility for considering the future in the light of the present by the aid of our scholarly methods. With our human economies precariously situated within natural forces that change us more than we can change them, energy is and has always been our very source of life; our culture is our self-defense. The Tsunami may not have been specifically climate-related or energy-conditioned in any meaningful way. However, it made us think again about the fragility of life, and our role. Within our societies, we need to consider carefully the prospects, costs and benefits of energy change, utilizing whatever conceptual instruments are available, not only those most ready at hand. What is sustainability, we have to keep asking, and how will we know when we have it?

The research reflected in this book points inward, to a deepening and broadening of perspectives already explored, and also outward to new horizons barely or never mentioned here. Already in the course of our research, we have traced variations in opinion, behaviour and legislation between the portions of the European area from north to south, from east to west. In this age of uniformization, unification and standardization, we may well wonder whether the most beneficial emerging patters among populations, governments and industries can possibly be subsumed under a single model. Can we impose our values and priorities even in our own societies – not to mention those outside the West – presuming we can agree what our values and priorities are?

Are our horizons, as scientists and as citizens, excessively oriented to our Western context? Energy saving and environmental responsibility are lower on the list of priorities in portions of the world where the fabric of society is torn asunder by endemic civil war or where the structures of government and economy are disrupted, nonexistent or arbitrary. Such concerns must rest in the background wherever policy success is measured exclusively by the possibility of survival from day to day. Yet the countries where priorities are different from ours have a voice among the 207 or so world nations; and they must be heard whenever the behaviour of the whole is discussed.

Further study would demand a closer look at relations between the West and the rest in an age of moral ambiguity. Will European and American foreign policy *vis-à-vis* oil rich areas of the developing world reflect concerns for global harmony also from an environmental standpoint? Will it reflect any agreed-upon concerns at all, apart from the logic of the boardroom? If not, how can the West demand from the developing world environmental disciplining according to the norms many of us have agreed to abide by? This is assuming of course that we in the West can indeed agree to abide by the very norms we have proposed.

We like to think of social science as having begun, some time in the last two centuries, as the product of a specific series of developments within Western civilization. It now belongs to everyone, and its practitioners across the globe may contribute to formulating the questions we need to ask and the research methods we need to explore. Widely varied as our separate social science disciplines may be, Max Weber defined the classic problem shared by all of them as follows: to analyse the scarcity of means. In modern times the problem is largely understood in terms of resources and availability, of which a considerable portion is contained within the relation between energy and culture. We are not the first to make this prediction, but let us state it in our own words: energy and culture will be the territory on which the struggle over the future of the planet will be carried out. On our ability to understand the ramifications of both of these terms rest the prospects for global harmony, and our hopes.

Index